Sustainable Civil Infrastructures

Sustainable Infrastructure impacts our well-being and day-to-day lives. The infrastructures we are building today will shape our lives tomorrow. The complex and diverse nature of the impacts due to weather extremes on transportation and civil infrastructures can be seen in our roadways, bridges, and buildings. Extreme summer temperatures, droughts, flash floods, and rising numbers of freeze-thaw cycles pose challenges for civil infrastructure and can endanger public safety. We constantly hear how civil infrastructures need constant attention, preservation, and upgrading. Such improvements and developments would obviously benefit from our desired book series that provide sustainable engineering materials and designs. The economic impact is huge and much research has been conducted worldwide. The future holds many opportunities, not only for researchers in a given country, but also for the worldwide field engineers who apply and implement these technologies. We believe that no approach can succeed if it does not unite the efforts of various engineering disciplines from all over the world under one umbrella to offer a beacon of modern solutions to the global infrastructure. Experts from the various engineering disciplines around the globe will participate in this series, including: Geotechnical, Geological, Geoscience, Petroleum, Structural, Transportation, Bridge, Infrastructure, Energy, Architectural, Chemical and Materials, and other related Engineering disciplines.

More information about this series at http://www.springer.com/series/15140

Deepankar Choudhury · Khalid M. El-Zahaby
Izzat Idriss
Editors

Dynamic Soil-Structure Interaction for Sustainable Infrastructures

Proceedings of the 2nd GeoMEast International Congress and Exhibition on Sustainable Civil Infrastructures, Egypt 2018 – The Official International Congress of the Soil-Structure Interaction Group in Egypt (SSIGE)

Springer

Editors
Deepankar Choudhury
Civil Engineering
Indian Institute of Technology Bombay
Mumbai, India

Izzat Idriss
University of California at Davis
California, CA, USA

Khalid M. El-Zahaby
Housing and Building National
Research Center
Giza, Egypt

ISSN 2366-3405 ISSN 2366-3413 (electronic)
Sustainable Civil Infrastructures
ISBN 978-3-030-01919-8 ISBN 978-3-030-01920-4 (eBook)
https://doi.org/10.1007/978-3-030-01920-4

Library of Congress Control Number: 2018957284

This Springer imprint is published by the registered company Springer Nature Switzerland AG
The registered company address is: Gewerbestrasse 11, 6330 Cham, Switzerland

Contents

About the Editors

Prof. Deepankar Choudhury, *FNASc*, IIT Bombay, Mumbai, India is Institute Chair Professor of Civil Engineering Department at Indian Institute of Technology (IIT) Bombay, Mumbai, India, and Adjunct Professor of Academy of Scientific and Innovative Research (AcSIR) of CSIR laboratories (connected to CSIR-CBRI, Roorkee) of India. Always first-class first-rank holder and gold medalist throughout academics in BE (gold medalist) from Jadavpur University Kolkata, India, in 1997, ME (gold medalist) from IISc Bangalore, India, in 2000, and PhD (with Prof. G. A. Leonard's Best PhD Thesis Award from Indian Geotechnical Society for years 2001 and 2002) from IISc Bangalore, India, in 2002, he also worked as a Faculty at IIT Kanpur, new IITs at Gandhinagar and Dharwad (both on deputation from IIT-B) and as a Visiting Fellow/Faculty at NUS, Singapore; UoW, Australia; UC Berkeley, USA; Kagoshima University, Japan; TU Darmstadt, Germany; and Incheon National University, South Korea. He is an elected Fellow (FNASc) of the oldest Science Academy of India, viz. The National Academy of Sciences, India. Internationally, he is Alexander von Humboldt Fellow of Germany, JSPS Fellow of Japan, and TWAS-VS Fellow of The World Academy of Sciences, Italy, in addition to National Fellow of The Institution of Engineers (India), Indian Geotechnical Society (IGS), Indian Society of Earthquake Technology (ISET). Prof. Choudhury, a recipient of Prof. C. S. Desai Medal of International Association for Computer Methods and Analysis in Geomechanics (IACMAG), USA, in 2017, also received Excellent Regional Contributions Award of IACMAG in 2014 and Excellent Paper Award (Junior) of IACMAG in 2008 for ASCE-International Journal of Geomechanics paper. He received several Young Scientist/Engineer Awards from various academies/societies of India like INSA, NASI, ISCA, INAE, IEI, DST, DAE, and ISTE, and from abroad like APACM, Shamsher Prakash Research Foundation, and IACMAG. He is an Associate Editor of ASCE-International Journal of Geomechanics, Indian Geotechnical Journal (Springer), Journal of The Institution of Engineers (India) - Series A (Springer), INAE Letters (Springer). He serves as Editorial Board Member for Canadian Geotechnical Journal, Geotechnical and Geological Engineering (Springer), and Geomechanics and Engineering (Techno-Press). His major research areas are civil geotechnical earthquake engineering, computational geomechanics, sustainable foundation engineering, and soil dynamics. Prof. Choudhury's video lectures in YouTube through NPTEL, MHRD, Govt. of India, on topics 'Soil Dynamics' and 'Geotechnical Earthquake Engineering' are highly popular all over the world. For his outstanding academic teaching and research, Prof. Choudhury received "Best Teacher" and "Best Researcher" awards from IIT Bombay. He already supervised 21 doctoral theses, some of which received Best PhD Thesis Awards and many more PhDs are continuing. His co-authored textbook on 'Foundation Systems for High-Rise Structures' is very popular, and he authored over 250 technical publications including over 130 peer-reviewed journal papers which are highly cited all over the world. He delivered several plenary/keynote/invited lectures at various reputed conferences all over the world. He is the current Secretary of Technical Committee TC 207 (Soil–Structure Interactions and Retaining Walls) of International Society for Soil Mechanics and Geotechnical Engineering (ISSMGE) and former Secretary of TC 212 (Deep Foundations) of ISSMGE, in addition to present and former member of TC 203 and Editorial Board of ISSMGE Bulletin. He executed several national and international projects of importance in India, Iraq, Mozambique, Vietnam, in the domain of his research expertise.

Prof. Dr. El-Zahaby graduated from the Faculty of Engineering, Cairo University, Egypt, in 1985 with honors. He obtained his MSc from Cairo University in 1989 and his PhD from Civil Engineering Department, North Carolina State University, Raleigh, NC, USA, in 1995. Upon graduation, he joined the Housing and Building National Research Center (HBRC) as a Research Assistant and was promoted to different positions. He is currently the HBRC Chairman. He is the recipient of the 2013 ICC Global Award of Atlantic City, NJ, USA, and has been a Geotechnical Consultant for almost 30 years. He is currently serving as an Assessor for Geotechnical Consultants in Egypt. He is also the Principal Investigator of a research project funded by the STDF, Egypt, and NSF, USA, on the assessment of seismic hazards on wind turbines.

I. M. Idriss ***Professional History:*** The Emeritus Professor of Civil Engineering at the University of California, Davis (UCD), and independent Consulting Geotechnical Engineer, currently residing in Santa Fe, New Mexico, USA. His areas of teaching, research, and practice are: geotechnical earthquake engineering; soil mechanics and foundation engineering; earthfill and rockfill dam engineering; and numerical modeling. He joined the faculty at UCD in 1989 following 20 years at Woodward-Clyde Consultants in San Francisco, Santa Ana, and Oakland where he was a Senior Principal. He was a Member of the teaching and research staff of the Geotechnical Engineering Group at the University of California at Berkeley from 1967 through 1975. He retired from the faculty position at UCD in July 2004.

Education and Registration: BCE, Rensselaer Polytechnic Institute in 1958; MS, Caltech in 1959; PhD, University of California at Berkeley in 1966. He is a registered Professional Engineer (PE) and Geotechnical Engineer (GE) in California.

Research and Consulting: He has conducted comprehensive post-earthquake investigations of some ten earthquakes since the 1964 Great Alaska earthquake. He has developed or co-developed many of the currently used procedures for evaluating the behavior of soil sites and soil structures during earthquakes.

He has been a consultant on many projects in the USA and other countries and serves on several consulting or review boards for USA and international companies and for State and US Federal Agencies. He has authored or co-authored about 250 technical papers and research reports on subjects related to the geotechnical aspects of earthquake engineering.

Awards and Honors: He has received many awards and honors over the past forty-five years, including election to the US National Academy of Engineering in 1989, receipt of the first H. Bolton Seed Medal from ASCE in 1995, the distinguished scholarly public service award from the University of California, Davis, in 1999, was elected an Honorary Member of the Japanese Geotechnical Society in 2005, was elected a Distinguished Member of ASCE in 2008, and is the recipient of the 2010 Ralph Peck Award from ASCE, and the 2010 Casagrande Memorial Lecture from the Boston Society of Civil Engineers. He was elected an Honorary Member of the Earthquake Engineering Research Institute (EERI) in 2012. He was the recipient of the George W. Housner Medal from EERI in 2018 and was selected to give the 2019 Terzaghi Lecture in 2019.

Effect of Embedment on Dynamic Response of Block Foundation Under Coupled Vibration

Kavita Tandon(✉), Rohit Ralli, Bappaditya Manna, G. V. Ramana, and Manoj Datta

Department of Civil Engineering, IIT Delhi, New Delhi, India
{Kavita.Tandon, cez158023, bmanna, ramana, mdatta}@civil.iitd.ac.in

Abstract. The dynamic response of a block foundation under machine-induced vibration depends on several parameters such as the shape and size of the foundation, depth of embedment, static load, dynamic excitation frequency and other geological parameters. In this present study, the dynamic response of embedded block foundation of aspect ratio $L/B = 1.5$ under coupled mode of vibration for different depth of embedment ($D = 0, 0.25$ and 0.5 m) has been studied. A number of block vibration tests are carried out using a Lazan-type mechanical oscillator under four different eccentric moments ($We = 0.2, 0.8, 1.4$ and 2.0 Nm) for a static load of 6.6 kN. The frequency amplitude responses of the block foundation for both translational and rotational motion for different embedded condition are also obtained using the half-space theory. The theoretical results are compared with the response obtained from the experimental investigations. The comparison showed encouraging agreement between the experimental test results and that obtained from the analysis. The frequency amplitude response of embedded block foundation indicates that as the embedment depth increases, the resonant amplitude decreases and resonant frequency increases.

Keywords: Dynamic response · Block foundation · Resonant amplitude Resonant frequency

1 Introduction

The design and prediction of soil-foundation system partially or fully embedded in the soil subjected to dynamic loads are of major importance in the field of soil dynamics. The dynamic response of such foundations is typically quantified in terms of various parameters that include geometrical and geological parameters such as the shape and size of the foundation, depth of embedment, dynamic excitation intensity, static load, soil mass density, shear modulus and damping ratio. Although most machine foundations are resting beneath the soil partially or fully, there exists little information about the effect of embedment on frequency amplitude responses of soil- foundation system. The effect of embedment of a circular footing under vertical vibration was first identified by Kaldjian (1971) using an elastic finite element solution. Novak and Beredugo (1972) developed an approximate analytical solution for embedded circular footings under coupled rocking and sliding motions. Novak and Sachs (1973) presented an

D. Choudhury et al. (Eds.): GeoMEast 2018, SUCI, pp. 1–7, 2019.
https://doi.org/10.1007/978-3-030-01920-4_1

approximate analytical solution for torsional vibrations of embedded footings partially embedded into a semi-infinite medium. Gazetas and Stokoe (1991) experimentally evaluated the dynamic response of model footings embedded at various depths. Gupta (1972) experimentally studied the effect of foundation embedment on resonant frequency, resonant amplitude, spring constant and damping factor of the soil-foundation system. Later, Gajan et al. (2005) presented experimental results of rotational stiffness degradation behaviour in terms of a mean rotational stiffness at a given rotation for a foundation rocking on sand. Pak et al. (2011) performed dynamic load test on scale model surface foundations and measured the vertical, horizontal and rocking response. The coupled vibration tests for different embedment depths reported in the literature are very limited. Therefore, in the present study experimental investigations are carried out for a model block foundation in the field to determine the frequency-amplitude response of soil-foundation system subjected to dynamic loads for both translational and rotational modes of vibration.

2 Experimentation

The block vibration tests for both translational and rotational modes of vibration were carried out on M30 grade concrete block of dimension 0.9 m × 0.6 m × 0.5 m for different eccentric moments (We = 0.221, 0.868, 1.450 and 1.944 Nm) with a static load (W_s) of 6.6 kN. The average density of soil in the field was 16.72 kN/m^3 with an average field N-value of 9 up to a depth of 3.0 m. The shear wave velocity (V_s) of the soil was found to be 216.42 m/s as obtained from the correlation (Hanumantharao et al. 2008) given in Eq. 1.

$$V_s = 86.0\,N^{0.42}\ \mathrm{m/sec} \tag{1}$$

Fig. 1. Experimental setup of block foundation under coupled horizontal and rocking vibration

The vibration tests were carried out for three embedment depths of $D = 0$, 0.25 and 0.5 m and time-acceleration response were measured using accelerometers connected with data acquisition system for different excitation frequencies. The complete experimental setup is shown in Fig. 1.

3 Elastic Half-Space Theory

The dynamic response of block foundations for surface footings is determined using the approximate analytical solution proposed by Velestos and Wei (1971). The methodology proposed by Velestos and Verbic (1973) and Novak and Beredugo (1972) for embedded foundations are used to calculate the stiffness and damping as available in the form of software DYNA 6. The theory considers soil as an elastic half-space and stiffness and damping constants of soil-foundation system are considered as frequency dependent. It is assumed that the dynamic reactions at the foundation base are equal to those of an elastic half space. The dynamic reactions acting at the foundation base are assumed to be equal to those of an overlying independent elastic layer.

4 Results and Discussions

4.1 Effect of Excitation Force Level

The typical frequency versus amplitude response curves obtained from the coupled horizontal and rocking vibration tests depicting the field testing and elastic half-space theory are shown in Figs. 2 (a) and (b) respectively. It can be seen from Figs. 2(a) and (b) that two resonant peaks occur at two different frequencies for both horizontal and rocking components. The second resonant peak is well separated from the first peak and is entirely suppressed. The coupled response is dominated by the first resonant peak of the horizontal translation as the rocking component is quite small. It is observed from the figures that as the excitation intensity increases the resonant frequency decreases about 2 to 10% and the resonant amplitude increases. The experimental resonant frequencies show a good qualitative agreement with the elastic half-space theory, whereas, the experimental resonant amplitudes are found to be considerably larger than the theoretical values.

4.2 Effect of Embedment Depth

The typical frequency versus amplitude curves of model block foundation with different embedment depths ($D = 0$, 0.25 and 0.5 m) subjected to coupled horizontal and rocking vibrations for both field testing and elastic half-space theory are shown in Figs. 3(a) and (b) respectively. It can be seen from the Figs. 3(a) and (b) that the maximum resonant amplitude of the block foundation decreases and the resonant frequency increases significantly with the increasing depth of embedment. It is observed from the experimental investigation that the decrease in resonant amplitude is about 20 to 35% for the embedment depth of 0.25 m and 30 to 50% for the embedment

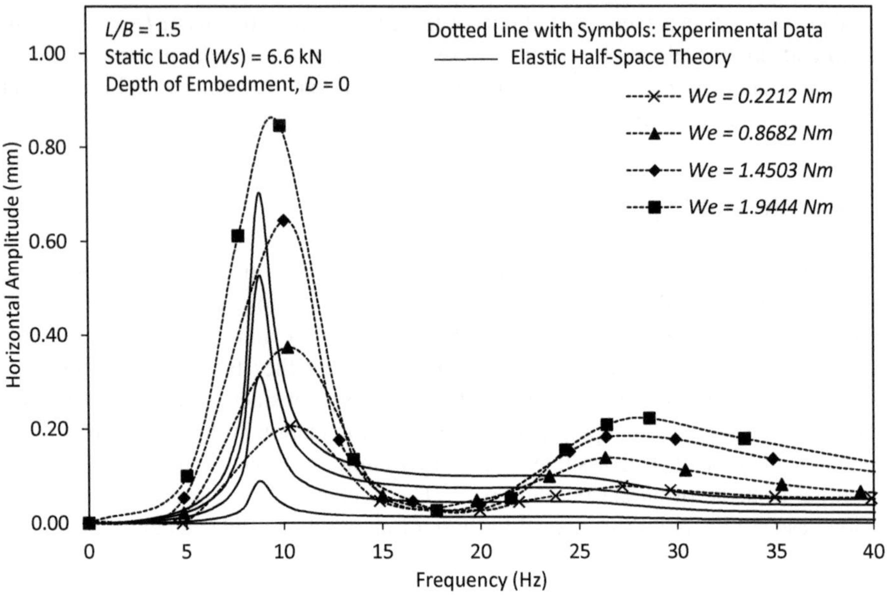

Fig. 2(a). Frequency versus amplitude curves of a surface block foundation under horizontal mode of vibration for different excitation intensities

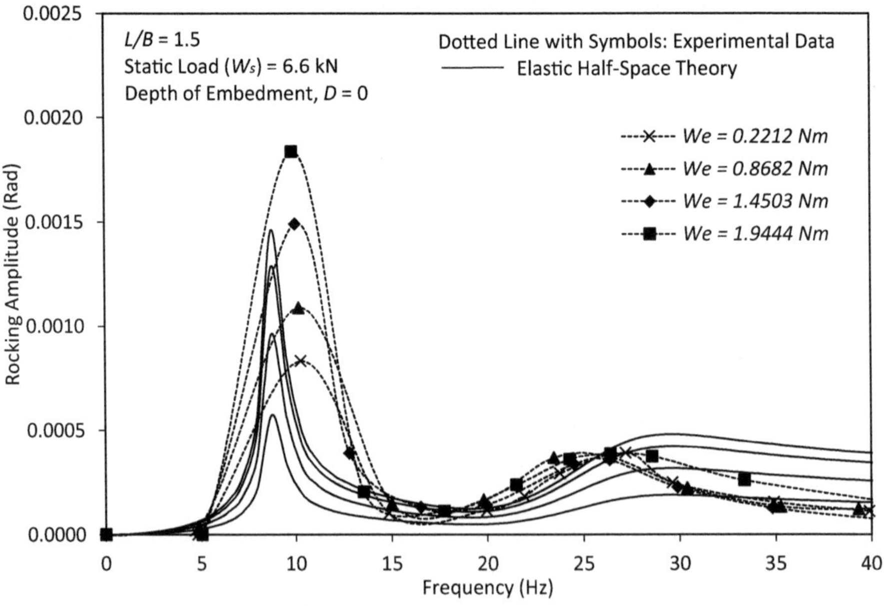

Fig. 2(b). Frequency versus amplitude curves of a surface block foundation under rocking mode of vibration for different excitation intensities

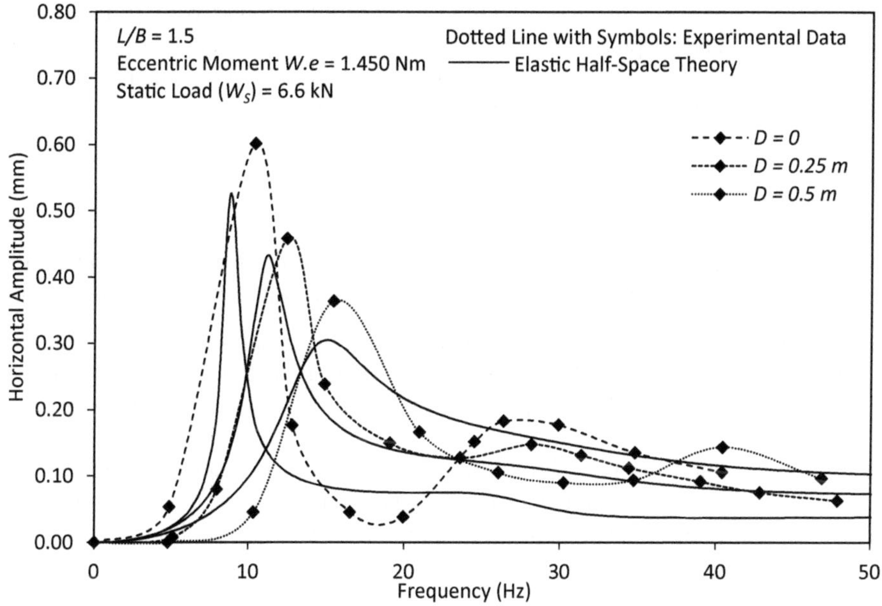

Fig. 3(a). Frequency versus amplitude curves under horizontal vibration for different embedment depths

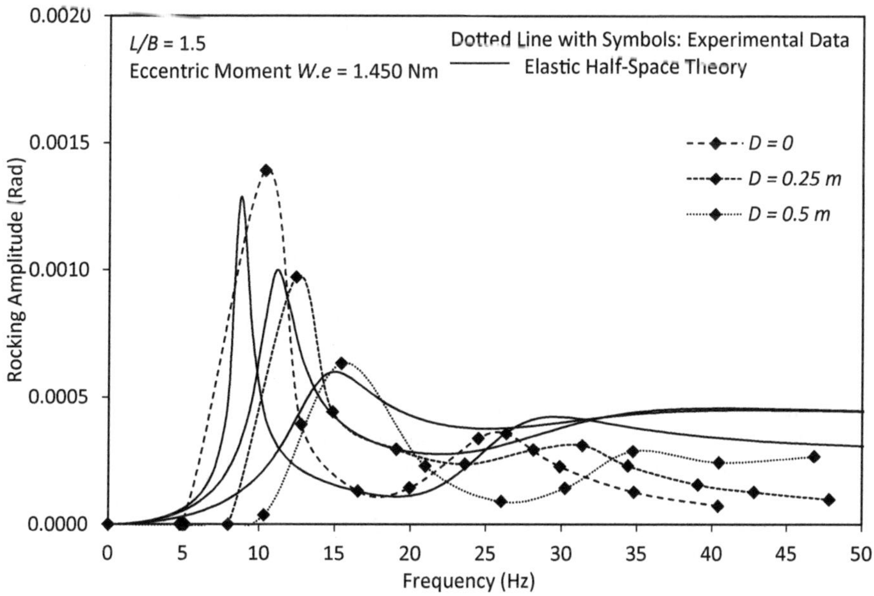

Fig. 3(b). Frequency versus amplitude curves under rocking vibration for different embedment depths

depth of 0.5 m, whereas, increase in resonant frequency is found to be 10 to 40% for different embedment depths as compared to surface block foundation. The experimental resonant frequencies and resonant amplitudes show a good qualitative agreement with the theoretical results for different embedment depths.

5 Conclusions

Based on the results of block vibration tests carried out on model block foundation in the field and analysis using elastic half-theory, the following conclusions are drawn:

1. Two resonant peaks are observed at two different frequencies for both translational and rotational components. The first resonant peak of the coupled response is dominated by the translation in horizontal direction as the rocking component is quite small.
2. The experimental resonant frequencies show a good qualitative agreement with elastic half-space theory for both translational and rotational modes of vibration. The experimental resonant amplitudes are found to be considerably larger than the theoretical values.
3. The resonant amplitude increases but the resonant frequency decreases with the increase in excitation intensity and both the first and second resonant frequencies decrease with increasing excitation intensity.
4. For a constant excitation level, the resonant amplitude decreases and resonant frequency increases with the increase in the embedment depth of foundation.
5. The increase in resonant frequency is about 10 to 40% whereas decrease in resonant amplitude is about 20 to 35% for the embedment depth of 0.25 m and 30 to 50% for the embedment depth of 0.5 m as compared to surface foundations.

References

Gajan, S., et al.: Centrifuge modelling of load-deformation behaviour of rocking shallow foundations. Soil Dyn. Earthq. Eng. (2005). https://doi.org/10.1016/j.soildyn.2004.11.019

Gazetas, G., Stokoe, K.H.: Free vibration of embedded foundations: theory versus experiment. J. Geotech. Eng. (1991). https://doi.org/10.1061/(asce)0733-9410(1991)117:9(1382)

Gupta, B.N.: Effect of foundation embedment on the dynamic behaviour of the foundation-soil system. Geotechnique (1972). https://doi.org/10.1680/geot.1972.22.1.129

Hanumantharao, C., Ramana, G.V.: Dynamic soil properties for Microzonation of Delhi. J. Earth Syst. Sci. (2008). https://doi.org/10.1007/s12040-008-0066-2

Kaldjian, M.J.: Torsional Stiffness of embedded footings. J. Soil Mech. Found. Eng. (1971)

Novak, M., Beredugo, Y.O.: Coupled horizontal and rocking vibration of embedded footings. Can. Geotech. J. (1972). https://doi.org/10.1139/t72-046

Novak, M., Sachs, K.: Torsional and coupled vibrations of embedded footings. Earthq. Eng. Struct. Dyn. (1973). https://doi.org/10.1002/eqe.4290020103

Pak, P.Y.S., et al.: Physical characteristics of dynamic vertical-horizontal-rocking response of surface foundations on cohesionless soils. Geotechnique (2011). https://doi.org/10.1680/geot.8.P.072

Velestos, A.S., Wei, Y.T.: Lateral and rocking vibration of footings. J. Soil Mech. Found. Eng. (1971)

Velestos, A.S., Verbic, B.: Vibration of viscoelastic foundation. J. Earthq. Eng. Struct. Dyn. (1973). https://doi.org/10.1002/eqe.4290020108

The Liquefaction Potential of Sandy Silt Layers Using the Correlation Between Penetrometer Test and SPT Test

Soukaina Touijrate[1(✉)], Khadija Baba[2], Mohamed Ahatri[1], and Lahcen Bahi[1]

[1] 3GIE Laboratory, Mohammadia Engineering School, Mohammed V University, Rabat, Morocco
stouijrate.st@gmail.com
[2] GCE Laboratory, High School of Technology-Salé, Mohammed V University, Rabat, Morocco

Abstract. Liquefaction is a hazardous and temporary phenomenon by which a soil saturated with water loses some or all of its resistance. The undrained conditions and a cyclic load increase the pores water pressure inside the soil and therefore a reduction of the effective stress.

Nowadays many semi-empirical methods are used to introduce a proposition to evaluate the liquefaction's potential using the in-situ test results. The objective of this paper is to study and compare the results of this semi-empirical methods and the numerical modelling results using the finite element methods.

The study is based on the data of the Pressuremeter test which be correlated to the Standard Penetrometer Test using the experimental results of the Casablanca-Tangier High-Speed Line exactly between PK 116 + 450 and PK 116 + 950 and near of Moulay Bousselham city. It belongs to the Drader- Soueir basin region which is located in the North-West of Morocco.

This region had a specific soil's formation, the first 50 m are characterized by the existence of sand layers alternating with layers of clay. These formations are very loose and saturated which suggests the possibility of soil liquefaction.

We present and discuss the results of applying the Idriss method and the Youd method in the evaluation of liquefaction susceptibility.

Apart from the previous empirical analysis to evaluate the liquefaction potential, numerical modelling is performed in this study.

1 Introduction

The soils, during liquefaction, behave as a liquid and no longer support the shearing forces, which causes large deformations which subsequently affect the structures built on this soil.

For assessing the liquefaction potential, many semi-empirical methods are developed. This potential reflects the soil ability to resist the cyclic shear efforts and is mainly dependent on the relative density, the particle size, the soil texture, the saturation degree and the earthquake magnitude.

D. Choudhury et al. (Eds.): GeoMEast 2018, SUCI, pp. 8–26, 2019.
https://doi.org/10.1007/978-3-030-01920-4_2

The case studied is based on the Pressuremeter test data for a section of the Casablanca-Tangier High-Speed Line exactly between PK 116 + 450 and PK 116 + 950, and that is part of the basin Drader-Soueir which is located in the North-West of Morocco.

The aim of this paper is, on the one hand, to evaluate and estimate the liquefaction potential using two semi-empirical methods, namely the Idriss method [12] and the Youd method [35], these methods are based on the SPT test data that we have already correlated from the Pressuremeter test data. On the other hand, to check and validate these correlation and methods using the finite element method.

The numerical modelling of the liquefaction behavior in this case is carried out in Quake/W software (from GeoStudio) [2], which simulates the soil's behavior under a cyclic loading by the generalized finite element method. This modelling will help us to determine the affinity of the correlation method and the semi-empirical method that predicts the best the liquefaction susceptibility.

2 Methods Used in the Evaluation

2.1 Correlation Method

The correlation study of Gonin et al. [11] is based on the comparison between the Standard Penetration Test results (SPT) and those of the Pressuremeter test on several geological formations: silts, sands, marls, limestone, …, in order to find the relations between the measured standard penetration resistance N and the pressure limits Pl and Pressurometric deformation modulus E_m of the Menard Pressuremeter test.

The study has helped to establish the correlations of the six main soil types where the test SPT and the Pressuremeter test are used and which are summarized in the Table 1.

Table 1. Simplified correlations proposed between N, Pl (MPa) and Em (MPa) [11]

Soil nature	Correlation Pl → N	Correlation Em → N
Silt	N = 32 × Pl	N = 2,6 × Em
Sands	N = 21 × Pl	N = 2,9 × Em
Greeb clays	N = 26 × Pl	N = 2,3 × Em
Plastic clays	N = 18 × Pl	N = 1,6 × Em
Marl	N = 23 × Pl	N = 1,9 × Em
Chalk	N = 6 × Pl	N = 0,7 × Em

2.2 Correction of the obtained value

The value of N obtained by the correlation corresponds to the number of shots measured on the site without any correction, which is influenced by several factors, the most important is the influence of the vertical effective stress.

This influence is translated by the C_N coefficient, knowing that:

$$N_{1,60} = C_N \times N_1 \tag{1}$$

Several formulas have been proposed for calculating the C_N coefficient, such as: Liao and Whitman [20]:

$$C_N = \left(\frac{P_a}{\sigma'_{v0}}\right)^{0,5} \tag{2}$$

Kayen et al. [17]:

$$C_N = \frac{2,2}{\left(1,2 + \frac{\sigma'_{v0}}{P_a}\right)} \tag{3}$$

Boulanger and Idriss [6]:

$$C_N = \left(\frac{P_a}{\sigma'_{v0}}\right)^{\alpha} \quad \text{With}: \ \alpha = 0,784 - 0,0768\sqrt{N_1} \tag{4}$$

The C_N cannot exceed the value of 1.7 [17, 24].

In our study, we used the Eq. (4) to determine the C_N value, since it is based on the results study of the first two correction formulas (2) and (3).

2.3 Semi-empirical Methods of the Liquefaction

The semi-empirical approaches evaluating the liquefaction potential are founded on the earthquakes observation, the liquefaction-induced effects, and the in-situ test results. The tests, which help to estimate the liquefaction potential, are mainly the Cone Penetration Test (CPT) and the Standard Penetration Test (SPT). These approaches can be divided into three categories:

- Approach by cyclic stress;
- Approach by cyclic deformation;
- Energy approach.

In this paper, the evaluation methods [12, 35] proposed are based on the cyclic stress approach using the SPT test data. This requires the calculation of the two main parameters in the estimating of the liquefaction potential which are the cyclic stress ratio CSR and the cyclic resistance ratio CCR in order to evaluating the safety coefficient.

The factor of safety is expressed by the same equation for all methods based on the cyclic stress approach [31]:

$$F_s = CCR/CSR \tag{5}$$

Evaluating cyclic stress ratio

The CSR estimate for the two methods studied is based on the simplified method of Seed and Idriss [31] which proposed the following relationship:

$$CSR = \frac{\tau_{av}}{\sigma'_{v0}} = 0,65\left(\frac{a_{max}}{g}\right)\left(\frac{\sigma_{v0}}{\sigma'_{v0}}\right)r_d \tag{6}$$

Where τ_{av} is the average cyclic shear stress; a_{max} is the maximum horizontal acceleration at the ground surface; g = 9,81 m^2/s is the acceleration of gravity; σ_{v0} is the initial vertical total stress; σ'_{v0} is the initial vertical effective stress; r_d is the stress reduction factor.

The stress reduction coefficient r_d is expressed in terms of the depth by the following equations (Liao and Whitman 1986) [20]:

$$\begin{aligned} r_d &= 1 - 0,00765 \qquad z \leq 9,15\,\text{m} \\ r_d &= 1 - 0,00765z \quad 9,15\,\text{m} < z \leq 23\,\text{m} \end{aligned} \tag{7}$$

This formula is used in engineering practice to estimate the average r_d, knowing that this value represents a wide range which increases with depth (Golesorki 1989) [10].

Also Seed and Idriss [31] proposed a mean and a range of rd values represented in Fig. 1 with the results of Eq. 6.

In Fig. 1 and for ease of calculation, the mean curve has been approximated by T. F. Blake (personal communication, 1996) by the following equation:

$$r_d = \frac{(1 - 0,4113z^{0,5} + 0,04052z + 0,001753z^{1,5})}{(1 - 0,4177z^{0,5} + 0,05729z - 0,006205z^{1,5} + 0,00121z^2)} \tag{8}$$

This equation is the one used by the Youd method [35] for determining r_d.

Idriss [14] is based on the Golesorkhi [10] work, found that r_d also depends on the earthquake ground motion characteristics (e.g. intensity and frequency), so it will be estimated according to the depth z and the earthquake magnitude M by the following formula:

$$\begin{aligned} &Ln(r_d) = \alpha(z) + \beta(z)M \qquad z \leq 34\,\text{m} \\ &r_d = 0,12\exp(0,22M) \qquad z > 34\,\text{m} \\ &\text{With}: \alpha(z) = -1,012 - 1,126\sin\left(\tfrac{z}{11,73} + 5,133\right) \\ &\beta(z) = 0,106 + 0,118\sin\left(\tfrac{z}{11,28} + 5,142\right) \end{aligned} \tag{9}$$

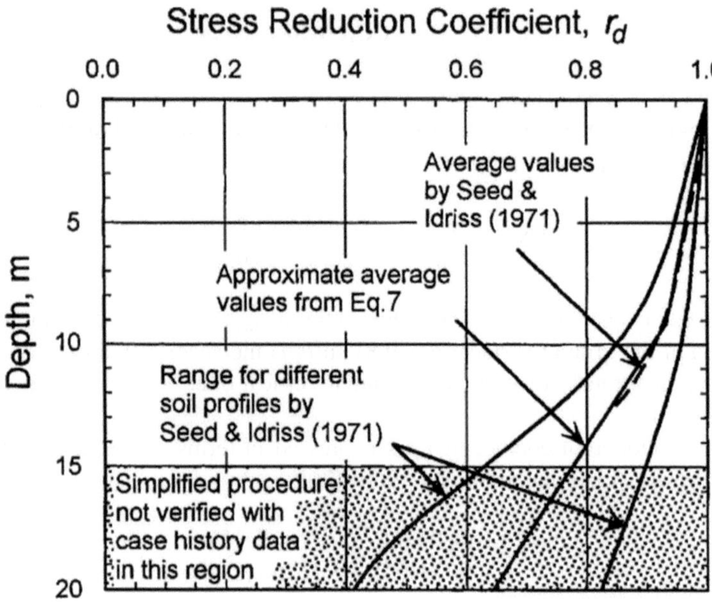

Fig. 1. r_d versus Depth Curves Developed by Seed and Idriss (1971) with Added

In the Idriss method [12] for the liquefaction potential estimating, the Eq. (6) has been adjusted by two new parameters, it becomes:

$$CSR = \frac{\tau_{av}}{\sigma'_{v0}} = 0,65\left(\frac{a_{max}}{g}\right)\left(\frac{\sigma_{v0}}{\sigma'_{v0}}\right)\frac{r_d}{MSF}\frac{1}{K_\sigma} \qquad (10)$$

Knowing that:

- MSF is the magnitude scaling factor, adjusts the CSR vis-à-vis the magnitude. it is expressed by Idriss [14] as:

$$MSF = 6,9\,exp\left(\frac{-M}{4}\right) - 0,058 \qquad \text{With : } MSF \leq 1,8 \qquad (11)$$

- K_σ is the overburden correction factor, adjusts the CSR vis-à-vis the initial vertical effective stress and the relative density D_r, it is expressed by Boulanger and Idriss [5], as:

$$K_\sigma = 1 - C_\sigma Ln\left(\frac{\sigma'_{v0}}{P_a}\right) \leq 1 \qquad \text{With : } \quad C_\sigma = \frac{1}{18.9 - 2.55\sqrt{N_{1,60}}} \leq 0.3 \qquad (12)$$

The $N_{1,60}$ is limited to maximum values of 37 [5].

Evaluating cyclic resistance ratio

The CRR evaluation in this study is based, directly, on the measured standard penetration resistance corrected of the SPT test.

Briefly, the CRR is expressed by the following equations:

- Idriss method [35]:

$$CRR = \exp\left\{\frac{N_{1,60}}{14.1} + \left(\frac{N_{1,60}}{126}\right)^2 - \left(\frac{N_{1,60}}{23.6}\right)^3 + \left(\frac{N_{1,60}}{25.4}\right)^4 - 2.8\right\} \quad (13)$$

- Youd method [12]:

$$CRR = \frac{1}{34 - N_{1,60}} + \frac{N_{1,60}}{35} + \frac{50}{\left(10N_{1,60} + 45\right)^2} - \frac{1}{200} \quad (14)$$

2.4 The Numerical Modeling

The numerical modeling of this work is done using Quake / W software which deals with the problematic by two analysis types: Initial in situ stress analysis and Dynamic analysis.

In a first place, we introduced the different materials properties for each layer in the surveys and the correlated N1,60 corrected and also specified the function of the PWP presented in Fig. 2.

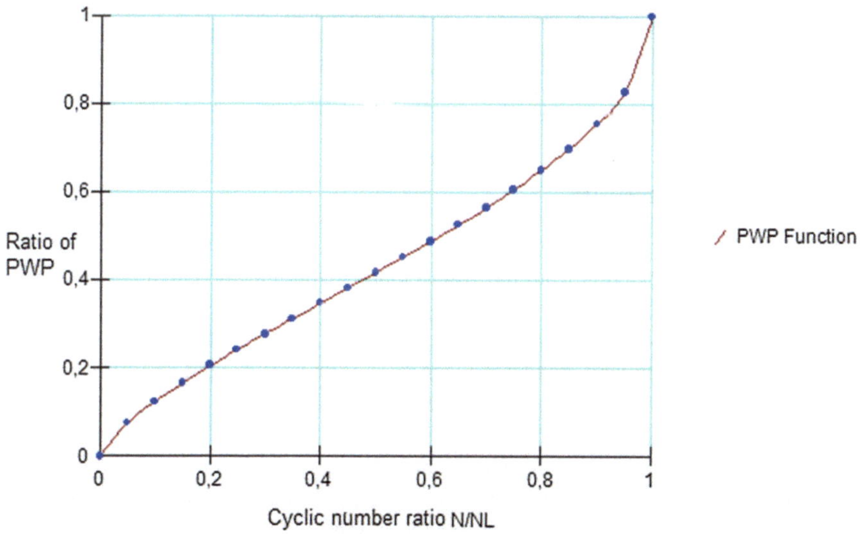

Fig. 2. PWP function

For the initial in situ stress analysis, we fixed X boundary conditions in the left and right boundaries and the XY boundary conditions in the bottom boundary.

For the dynamic analysis, we fixed Y boundary conditions in the left and right boundaries and the XY boundary conditions in the bottom boundary.

The mesh size is 1 m.

3 Case Study

Our case study positioned at the southern section of the High-Speed Line linking Tangier and Casablanca, it is between the PK 116 + 450 and the PK 116 + 950. Geographically, this section is located in the North-West of Morocco, precisely, 65 km South-East of Moulay Bousselham city (Fig. 3).

Fig. 3. Satellite view of the section studied

3.1 Geological Framework

Geologically (Fig. 4), the section studied is part of the Drader-Soueir basin, which belongs to a unit that extends from the Gharb plain's South to the Oued Loukkos in the North and is limited by the clay and marly pre-Rif outcrops in the East. The unit is composed of the plio-quaternary soils that are quite varied but always more or less detrital (intercalation of sand, sandstone, sandstone limestone, sandy clay) resting on a very clear substratum formed by the Miocene blue marls [29].

The soils in the area are profoundly influenced by the textural nature of their bedrock made of quaternary sands. The granulometric analysis shows that the majority of the soils in this area contain at least 75% of the total sands, the most often 90%. The clay contents are located in the 4 to 6% except in depth where they sometimes reach 18 to 20%. The deep horizon frequently includes calcareous kidneys and needles.

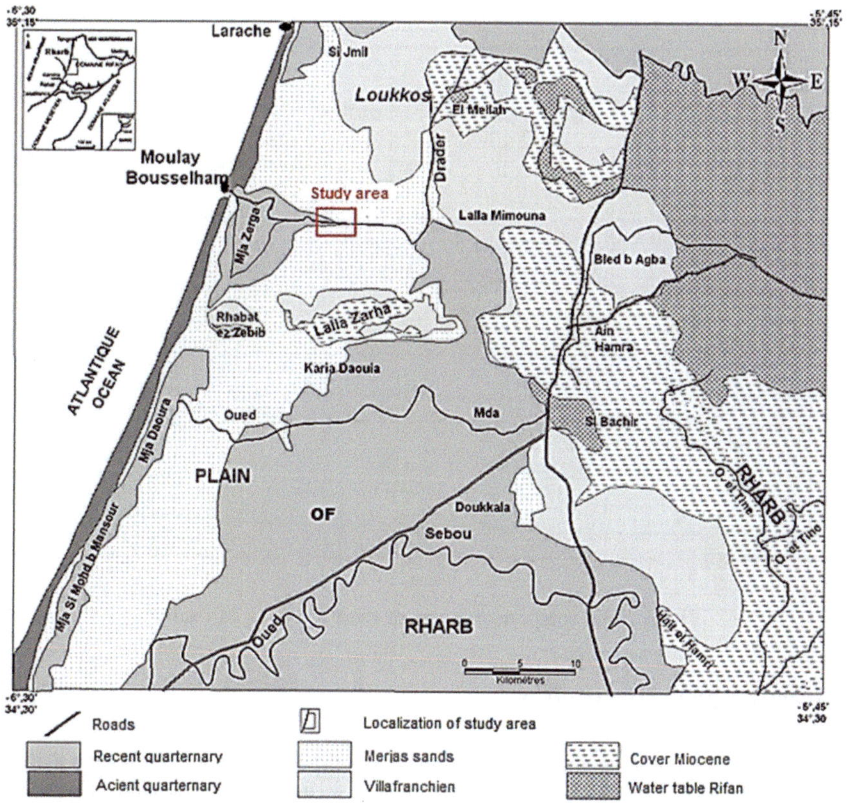

Fig. 4. The Pre-Rif Simplified Geological Map [29] Mean-Value Lines Plotted from Eq. (7) [31]

3.2 The Region's Seismicity

In Oued Drader, the maximum intensity is measured at near VII, but due to the proximity of the area to the critical seismic intensity regions, including a historical seismic site, and for safety measures, considers an intensity of IX.

Many conversion laws between the intensity and the magnitude were developed by Cherkaoui [33] and Karnick [16], thus allowing a correlation between the two parameters. For an intensity equal to IX, the Drader-Soueir basin magnitude retained is M = 6.5.

Given that the Oued Drader region is part of the Seismic Zone No. 3 according to the RPS 2000 revised in 2011, it is therefore considered a maximum acceleration of 0.14 g [19].

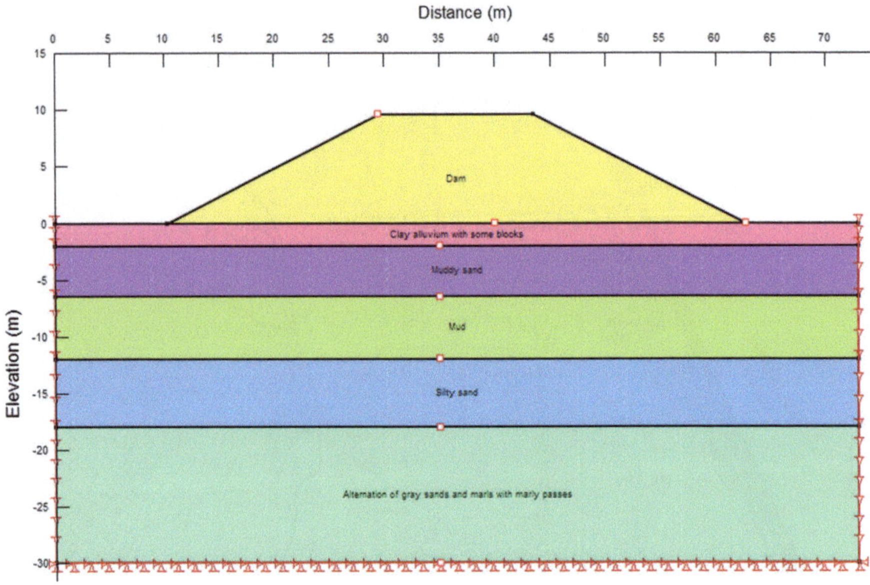

Fig. 5. The component layers of the PK 116 + 746 survey

4 Results and Discussions

In this study, we evaluate the correlation and the liquefaction susceptibility using the data from 2 surveys of Pressuremeter test located in the section between the PK116 + 450 and the PK 116 + 950 of the High-Speed Line, which are the PK116 + 746 (Fig. 5) and the PK116 + 900 (Fig. 6).

4.1 The Correlation Comparison

Using the Table 1 [11], we correlated the Pressuremeter test data from the two surveys cited above.

The correlation by the GONIN et al. method [11], is done using the parameters of the Pressuremeter test Pl and Em each of which gives a different N1 value, these values are presented in Figs. 7 and 8.

For the first survey PK116 + 746, the difference between the N1 values correlated from Pl and Em are lower for the first 3 layers but starts to be higher for the other two layers.

This is the same case for the second survey PK116 + 900 except for the grey marl layer.

This difference is not related to soil type since we found a large difference between the layers results which have the same grain size.

By contrast, we note that the difference between N1-Pl and N1-Em increases for the two surveys in a similar depth order which is 12 m for the survey PK116 + 746 and 14 m for the survey PK116 + 900.

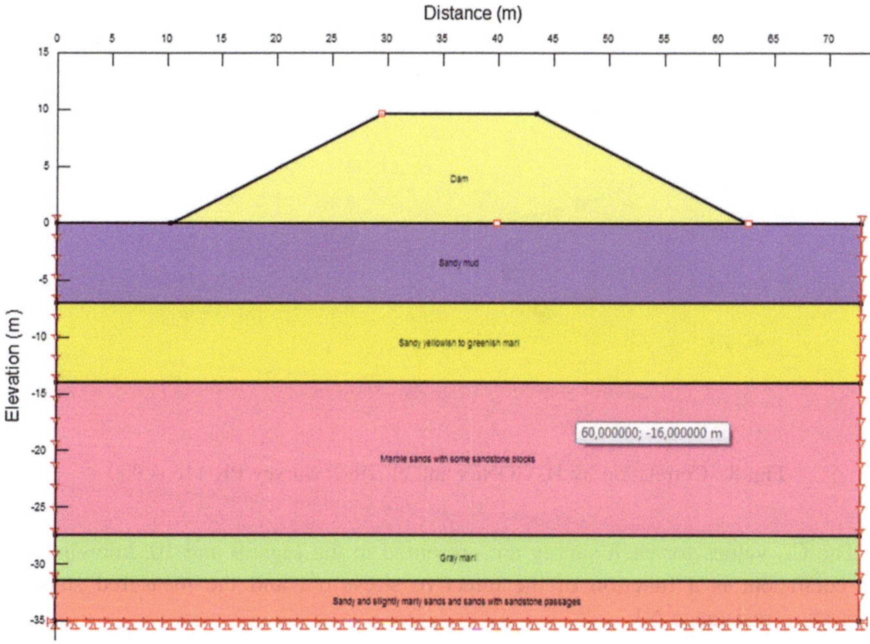

Fig. 6. The component layers of the PK 116 + 746 survey

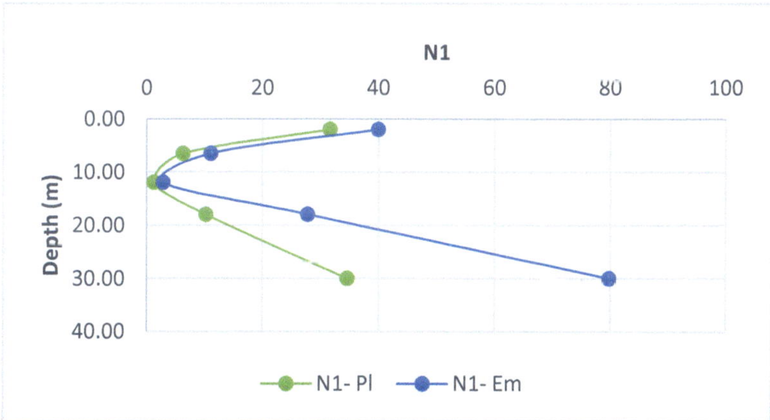

Fig. 7. Correlation by H. GONIN and al. 1992: Survey PK 116 + 746

4.2 The Correction of the Obtained Value for Each Survey

The values obtained are corrected by the coefficient C_N of the Eq. 4, which presents the vertical effective stress influence.

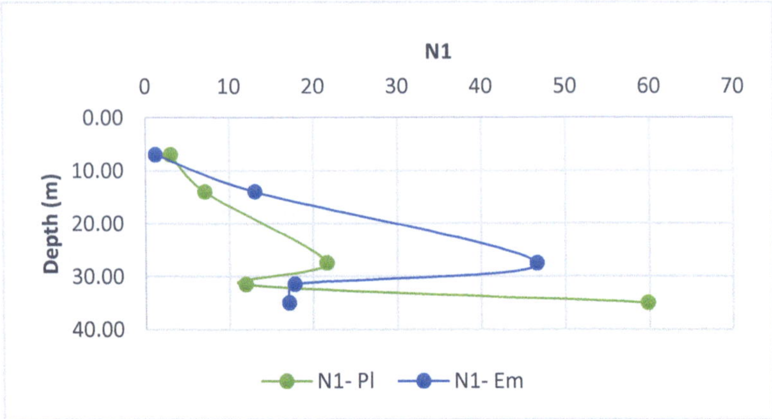

Fig. 8. Correlation by H. GONIN and al. 1992: Survey PK 116 + 900

The C_N values for each survey are presented in the Figs. 9 and 10, knowing that this coefficient is a function of the effective stress σ'_{v0} and the measured standard penetration resistance N1.

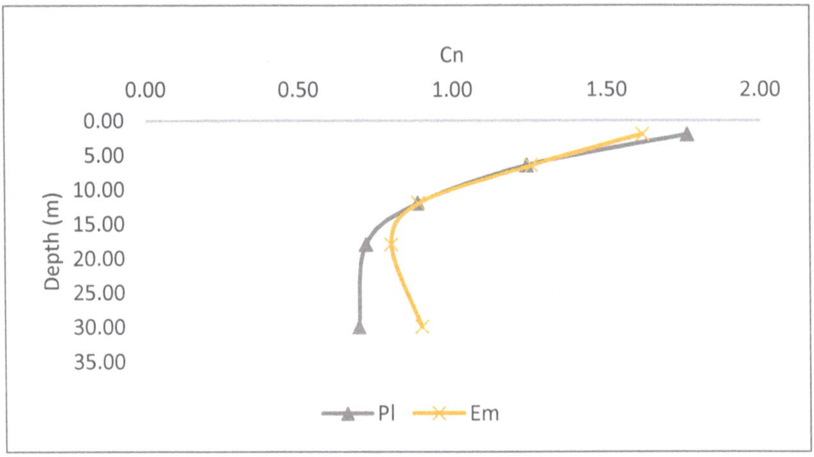

Fig. 9. The coefficient of the vertical effective stress influence: Survey 116 + s746

The graphs show that the C_N coefficient calculated by the N1-Pl and N1-Em are almost similar in the first 15 m, but the difference between them becomes larger in more significant in the deeper depths for each survey.

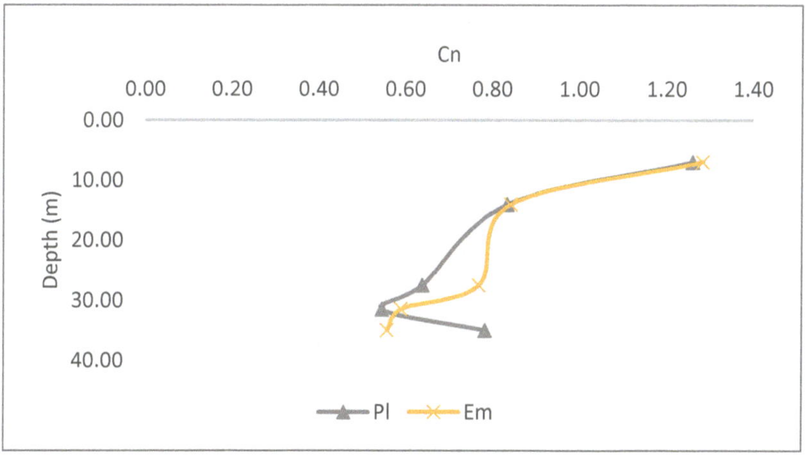

Fig. 10. The coefficient of the vertical effective stress influence: Survey 116 + 900

4.3 Validation and Verification by the Numerical Modelling

In order to evaluate, on the one hand, the correlation between the Pressuremeter test and the SPT test and, on the other hand, the semi-empirical methods results for estimating the liquefaction potential, we have modeled the two surveys PK116 + 746 and PK116 + 900, using the correlated data of the pressures limits as well as that of the Pressuremeter module.

Correlations

Comparing the modeling results (Figs. 11 and 12) with the two semi-empirical methods results (Figs. 13 and 14) based on the correlated data from the Pressuremeter modules, we found a large difference for each of the surveys and in each estimation method.

By against, a similarity is remarkable between the modeling results (Figs. 15 and 16) and semi empirical methods results (Figs. 13 and 14) based on the correlated data from the pressure limits.

Semi-empirical methods for evaluating the liquefaction potential

The comparison and the verification of the semi-empirical methods is based on the SPT data correlated from the pressure limit data of the Pressuremeter test in each surveys PK116 + 746 and PK116 + 900.

From Figs. 13 and 14, we observe that the curves of the two methods have the same allure with a clear gap mainly due to the introduction of the adjustment coefficients MSF and K_σ and the different formulas of the stress reduction coefficient r_d.

The MSF and K_σ coefficients have adjusted the CSR values (Figs. 17 and 18) of the Idriss method [12], making them almost constant. On the other hand, the CSR values of the Youd method [35] vary for each layer.

These CSR values are also influenced by the r_d (Figs. 19 and 20) whose allures for each method depend essentially on its equation as a function of depth. In Figs. 19 and 20, we observe that the r_d values of the Idriss method [12] are always less than those of Youd method [35].

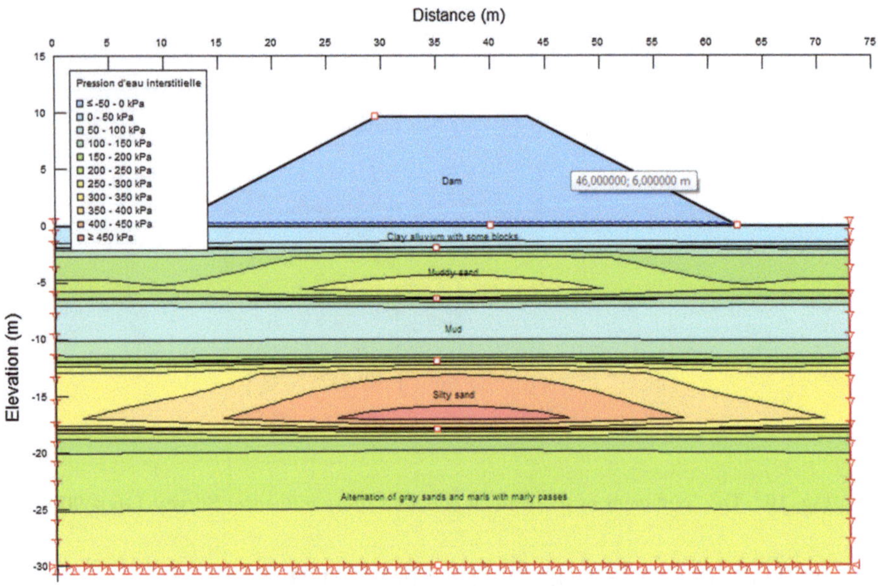

Fig. 11. The PWP distribution in the PK 116 + 746 survey – Em correlated data

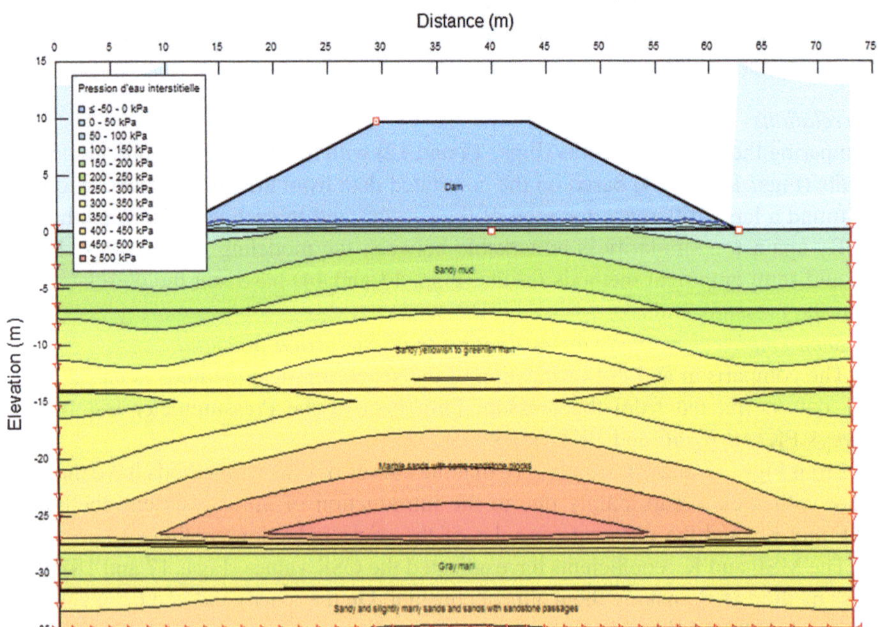

Fig. 12. The PWP distribution in the PK 116 + 900 survey – Em correlated data

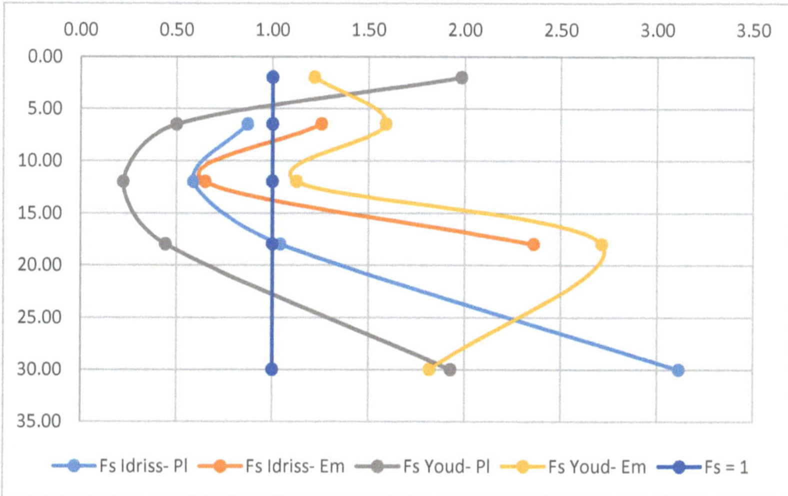

Fig. 13. The Fs values in the PK 116+746 survey

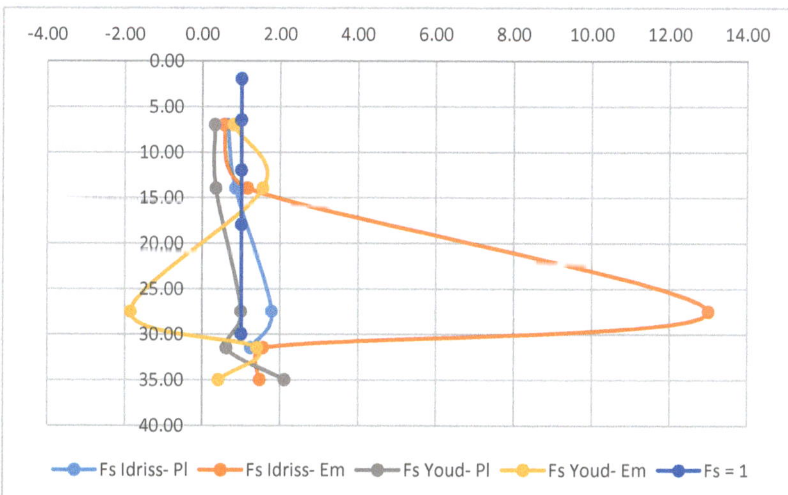

Fig. 14. The Fs values in the PK 116+900 survey

The Idriss method [12] supposes several corrections, moreover, when the N1,60 exceeds 37 blows, it can no longer obtain the safety coefficients. And comparing it with the modeling results (Figs. 15 and 16), we observe that it does not predict all the points likely to liquefy.

For Youd method [35], it is easier to operate. In comparison with the modeling (Figs. 15 and 16), we found that it gives almost all the points likely to be liquefied but it is binding for the muddy and marly layers.

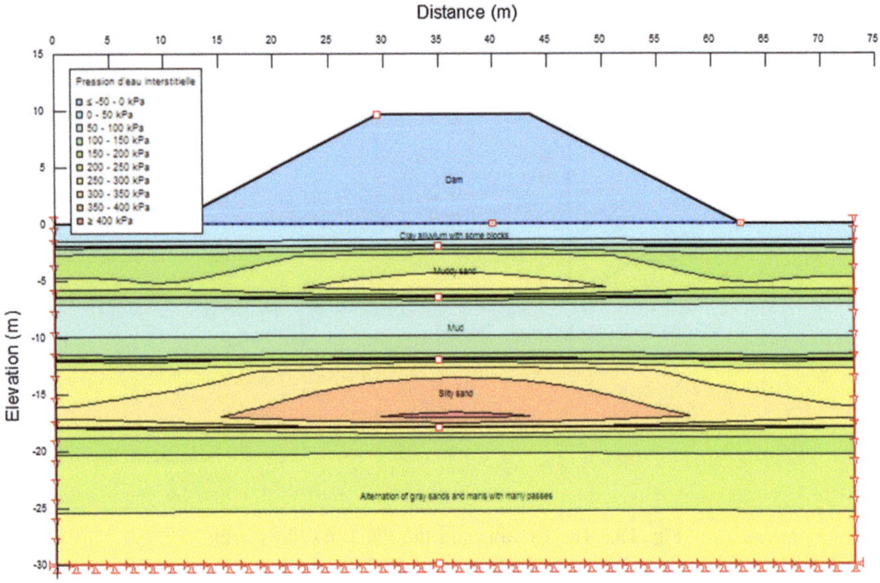

Fig. 15. The PWP distribution in the PK 116 + 746 survey – Pl correlated data

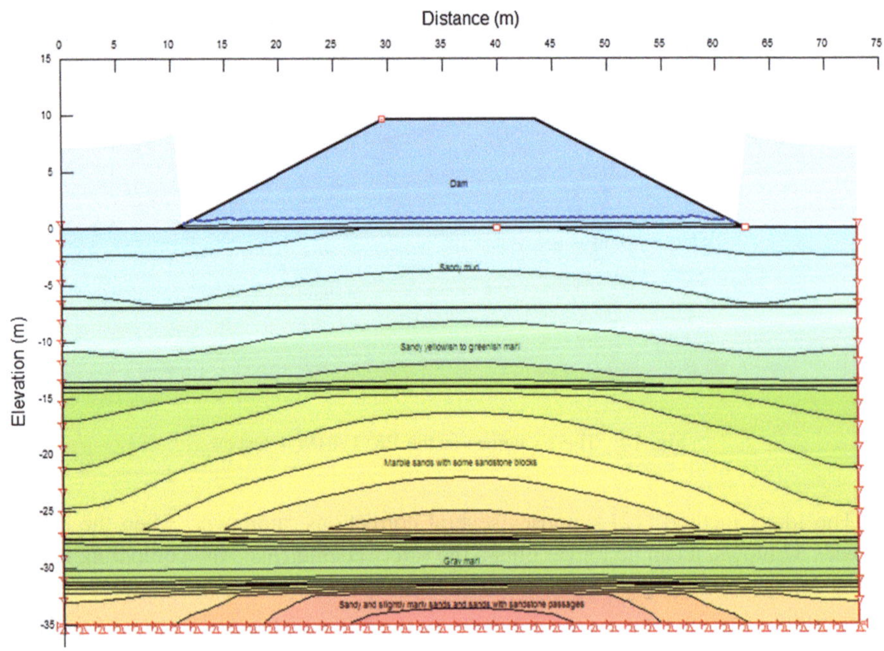

Fig. 16. The PWP distribution in the PK 116 + 900 survey – Pl correlated data

Fig. 17. The CSR values in the PK 116 + 746 survey

Fig. 18. The CSR values in the PK 116 + 900 survey

Evaluating liquefaction potential

On the basis of the surveys modeling results (Figs. 15 and 16) sustained by the semi-empirical methods results (Figs. 13 and 14), we found that the layers that are highly susceptible to liquefy are "silty sand", "Marble sands with some sandstone blocks" and "Sandy and slightly marly sands and sands with sandstone passages" layers and the moderately likely are the "Muddy sands" and "alternation of gray sands and marls with marly passes".

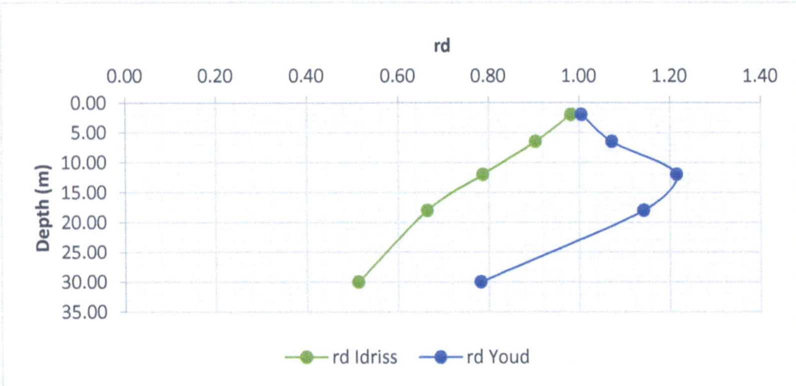

Fig. 19. The stress reduction coefficient r_d in the PK 116 + 746 survey

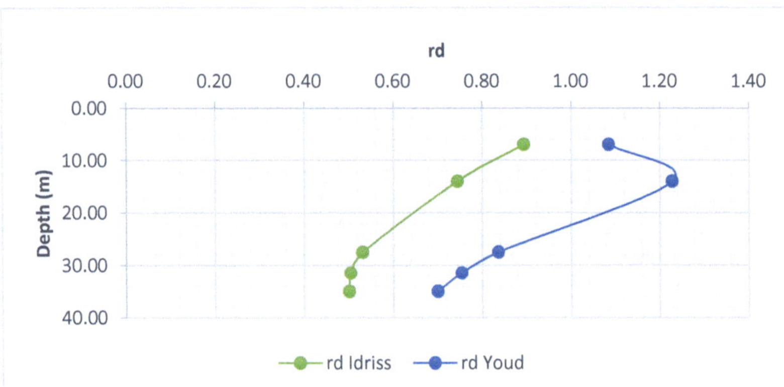

Fig. 20. The stress reduction coefficient r_d in the PK 116 + 900 survey

5 Conclusions

To conclude, the correlation between the Pressuremeter test and the SPT test from the pressures limits is more accurate than that of the Pressuremeter module.

For the semi-empirical methods of evaluating the liquefaction potential, the Idriss method [12] is limited by the blows number and is less secure compared to the Youd method [35], even though the latter is binding for the muddy and marly layers.

And finally, the Drader region presents a high risk of the liquefaction especially in the "silty sand" and "Sandy and slightly marly sands with sandstone passages" layers.

References

1. Pecker, A.: Livre de Dynamique des sols, Presse de l'école nationale des Ponts et Chaussées (1984)
2. An Introductory QUAKE/W Example, GEO-SLOPE International Ltd, Calgary, Alberta, Canada. www.geo-slope.com
3. Borowiec, A., Maciejewski, K.: Assessment of Susceptibility to Liquefaction of Saturated Road Embankment Subjected to Dynamic Loads Studia Geotechnica et Mechanica, Vol. XXXVI, No. 1 (2014) https://doi.org/10.2478/sgem-2014-0002
4. Arab, A., Sadek, M., Shahrour, I.: Influence de la densité relative sur la résistance au cisaillement cyclique des sables. In: MATEC Web of Conferences 149, 02034, CMSS-2017 (2018). https://doi.org/10.1051/matecconf/201814902034
5. Boulanger, R.W., Idriss, I.M.: State normalization of penetration resistance and the effect of overburden stress on liquefaction resistance. In: Proceedings, 11th International Conference on Soil Dynamics and Earthquake Engineering and 3rd International Conference on Earthquake Geotechnical Engineering, University of California, Berkeley, CA (2004)
6. Boulanger, R.W.: High overburden stress effects in liquefaction analyses. J. Geotech. Geoenviron. Eng. ASCE **129**(12), 1071–1082 (2003)
7. Castelli, F., Cavallaro, A., Grasso, S., Maugeri, M.: Soil liquefaction and risk analysis from in situ tests for the city of Trapani (Italy). In: Fourth International Conference on Recent Advances in Geotechnical Earthquake Engineering and Soil Dynamics (2001)
8. Castelli, F., Lentini, V.: SPT-based evaluation of soil liquefaction risk. In: Fifth International Conference on Recent Advances in Geotechnical Earthquake Engineering and Soil Dynamics (2010)
9. Zhang, G., Robertson, P.K., Brachman, R.W.I.: Estimating liquefaction-induced lateral displacements using the standard penetration test or cone penetration test, Project Engineer, EBA Engineering Consultants Ltd., 14940-123 Ave., Edmonton AB, Canada T5 V 1B4. J. Geotech. Geoenvironmental Eng. © ASCE (2004)
10. Golesorkhi, R.: Factors influencing the computational determination of earthquake-induced shear stresses in sandy soils. Ph.D. thesis, University of California, Berkeley, 395 p. (1989)
11. Gonin, H., Vandangeon, P., Lafeuillade, M.-P.: Etude sur les corrélations entre le standard et le pressiomètre penetration test, Rev. Franç. Géotech. ho 58, pp. 67–78 fianvier L992
12. Idriss, I.M., Boulanger, R.W.: Semi-empirical procedures for evaluating liquefaction potential during earthquakes. Soil Dyn. Earthq. Eng. **26**(2006), 115–130 (2004)
13. Idriss, I., Sun, J.: User Manual for SHAKE 91. Technical report, Center for Geotechnical Engineering Modeling, Departament of Civil Engineering, University of California, Davis. U.S.A (1992)
14. Idriss, I.M.: An update to the Seed-Idriss simplified procedure for evaluating liquefaction potential. In: Proceedings, TRB Workshop on New Approaches to Liquefaction, January, Publication No. FHWA-RD-99- 165. Federal Highway Administration (1999)
15. Andrade, J.E.: A predictive framework for liquefaction instability. Géotechnique **59**(8), 673–682 (2009). https://doi.org/10.1680/geot.7.00087
16. Karnik, V.: Seismicity of the European area. Part I., Prague and Dordrecht-Holland, 364 p. (1969)
17. Kayen, R.E., Mitchell, J.K., Seed, R.B., Lodge, A., Nishio, S., Coutinho, R.: Evaluation of SPT-, CPT-, and shear wave-based methods for liquefaction potential assessment using Loma Prieta data. In: Proceedings, 4th Japan-U.S. Workshop on Earthquake-Resistant Des. of Lifeline Fac. and Countermeasures for Soil Liquefaction, Vol. 1, pp. 177–204 (1992)

18. Kramer, S.L.: Geotechnical Earthquake Engineering Prentice, Hall Upper Saddle River, New Jersey, pp. 274–275 (1996)
19. Le règlement de construction parasismique 2000-version 2011, Ministre de l'habitat et la politique de la ville
20. Liao, S.S.C., Whitman, R.V.: A catalog of liquefaction and Non-liquefaction occurrences During Earthquakes, Department of civil Engineering, Massachusetts institute of technology, Cambridge, MA, 117 p. (1986)
21. Matasovic, N.: Seismic response of composite horizontally-layered soil deposits. Ph.D. thesis, University of California, Los Angeles (1992)
22. Bensoula, M., Missoum, H., Bendani, K.: Critical undrained shear strength of loose-medium sand-silt mixtures under monotonic loadings. J. Theor. Appl. Mech. **53**(2), 331–344 (2015). https://doi.org/10.15632/jtam-pl.53.2.331
23. Rahhal, M.E.: Comprendre les méthodes d'évaluation du potentiel de liquéfaction des sols. In: Proceedings of the 4th Canadian Conference on Geohazards: From Causes to Management. Presse de l'Université Laval, Québec, 594 p. (2008)
24. National Center for Earthquake Engineering Research (NCEER). In: Youd, T.L., Idriss, I.M. (eds.) Proceedings of the NCEER Workshop on Evaluation of Liquefaction Resistance of Soils, Technical report NCEER-97-022 (1997)
25. Newmark, N.M.: A method of computation for structural dynamics. J. Eng. Mech. Div. **85**, 67–94 (1959)
26. Touil, N.: Thèse sous le thème «Analyse expérimentale et numérique du potentiel de liquéfaction des sols sableux de Tanger» Université ABDELMALEK SAADI (2012)
27. Yagiz, S., Akyol, E., Sen, G.: Relationship between the standard penetration test and the pressuremeter test on sandy silty clays: a case study from Denizli. Bull. Eng. Geol. Environ. **67**, 405–410 (2008). https://doi.org/10.1007/s10064-008-0153-2
28. Ahmad, S., Khan, M.Z., Anwar, A., Mohd, S., Husain, A.: Assessment of Liquefaction Potential of Cohesionless Soil by Semi-Empirical: SPT- Based Procedure, ISSN (Online), vol. 3, Issue 10, pp. 2347–2812 (2015)
29. Maychou, S.: Étude morphostructurale et cartographie SIG du Rharb Septentrional et du Prérif (Maroc). Analyse sismotectonique et modélisation de la déformation de la région de Moulay Bousselham, 2009, l'Université Chouaïb Doukkali et L'Université Bordeaux 1
30. Seed, B., Idriss, I.M.: Ground motions and soils liquefaction during earthquakes. Technical report, Earthquake Engineering Research Institute (1982)
31. Seed, H.B., Idriss, I.M.: Simplified procedure for evaluating soil Liquefaction potential. J. Soil Mech. Found. Div. ASCE **97**(SM9), 1249–1273 (1971)
32. Touijrate, S., Baba, K., Ahatri, M., Bahi, L.: Validation and verification of semi-empirical methods for evaluating liquefaction using finite element method. In: MATEC Web of Conferences 149, CMSS-2017 02028 (2018). https://doi.org/10.1051/matecconf/201814902028
33. Cherkaoui, T.-E., Asebriy, L.: Le risque sismique dans le Nord du Maroc, Trav. Inst. Sci. Rabat, sér. Géol. & Géogr. phys., n° 21, pp. 225–232 (2003)
34. Terzaghi et Peck: Mécanique des sols appliquée aux travaux publics et aux bâtiments. Dunod, Paris (1961)
35. Youd, T., Idriss, I.M., Andrus, R.D., Arango, I., Castro, G., Chistian, J.T.: Liquefaction resistance of soils: summary report from the 1996 NCEER and 1998 NCEER/NSF workshop on evaluation of liquefaction resistance of soils. J. Geotech. Geoenvironmental Eng. ASCE, **127**(10) (2001)

Seismic Soil Structure Interaction Analysis of Rigid Piled Isolated Footing for Mid Rise Building in a Weak Soil

Pallavi Badry[1(✉)] and Ravi Shankar Badry[2]

[1] Vidya Jyothi Institute of Technology, Hyderabad, India
pallavibadrim@vjit.ac.in
[2] ARUP India Pvt Ltd, Hyderabad, India
ravishankar.badry@arup.com

Abstract. The piled raft foundation is generally a recommended choice of foundation for the building founded in a weak strata. To analyze the system in an accurate manner, it is needed to incorporate the interaction effect at the soil foundation interface. The direct method of interaction analysis facilitates to model the foundation system and the superstructure at integrated manner and thus models the system in a more realistic facet. In this study, the attempt is made to investigate the performance of isolated footing associated with rigid piles as a choice of foundation for mid rise building with a weak local soil condition. The interaction effect is modeled and the integrated model of the superstructure and foundation system is analyzed against static forces. The performance of the rigid piled isolated footing is compared in terms of system displacement and stresses generated with the piled raft (separately) foundation for same superstructure and different soil conditions. The proposed foundation type is found to be commensurate and economical with the set of limitations.

1 Introduction

All engineered construction resting on the soil, founded on a load transferring interfacing element called a foundation. The foundation is the part of an engineered system that transmits to, and into, the underlying soil or rock the loads supported by the foundation and its self-weight. Thus, type and efficiency of the load carrying mechanism are highly important and thus needed to design an accurate manner. There are several types of foundation systems, including isolated footing, strip footing, spread footing, pie foundation, cassions and wells which are recommended depending upon the type of soil and load for which system is needed to be designed. The following Table 1 gives the overall idea of the types of foundation and its recommendation for the typical soil condition.

D. Choudhury et al. (Eds.): GeoMEast 2018, SUCI, pp. 27–39, 2019.
https://doi.org/10.1007/978-3-030-01920-4_3

Table 1. Foundation types and reccommandation (BS 8004)

Foundation type	Use	Applicable soil conditions
Shallow foundations (generally D/B $\leq$ 1)		
Spread footings, wall footings	Individual columns, walls	Any conditions where bearing capacity is adequate for applied load. May use on a single stratum; firm layer over soft layer or soft layer over firm layer. Check settlements from any source
Combined footings	Two to four columns on footing and/or space is limited	Same as for spread footings above
Mat foundations	Several rows of parallel columns; heavy columns loads; use to reduce differential settlements	Soil bearing capacity is generally less than for spread footings, and over half the plan area would be covered by spread footings. Check settlements from any source.
Deep foundations (generally D/B $\geq$ 4)		
Floating pile	In groups of 2^+ supporting a cap that interfaces with columns(S)	Surface and near –surface soils have low bearing capacity and competent soil is at great depth. Sufficient skin resistance can be developed by soil-to- pile perimeter to carry anticipated loads.
Bearing pile	Same as for floating pile	Surface and near –surface soils not relied on for skin resistance; competent soil for point load is at a practical depth (8-20 m). Same as for piles may be floating or point-bearing (or combination).
Drilled piers or caissons	Same as for piles; use fewer; for large column loads	Depends on depth to competent bearing stratum.

2 Literature Survey

Several researchers contributed to understanding the bearing capacity of foundation and its variation wrt soil type and foundation geometry. The seismic analysis of the foundation and its bearing capacity and permanent ground deformation is studied for different superstructure configuration Debasis Roy (Roy 2013). The seismic pseudo-static analysis for retaining wall foundation with soil foundation interaction is carried out to evaluate the ultimate bearing capacity factors are computed for various values of soil friction angle, seismic acceleration coefficients in horizontal and vertical directions, ground inclination, and distance of the foundation from the slope edge with lateral earth pressure consideration (Ghazavi and Mahali 2014).

The bearing capacity variant of the small to medium rise buildings, shallow foundations is frequently used in hilly regions and its significance is studied and found bearing capacity and building configuration impacts the response of the system (Dhiraj and Bharathi 2014). Sahoo (2013) discussed about the bearing capacity and settlement study of shallow footings is a subject which needs consideration for the design of a foundation and concluded that the increase of stress in soil layers due to the load imposed by various structures at the foundation level will always be accompanied by some strain, which will result in the settlement of the structures. Srilakshmi and Rekha (2011) studied three possible failure mechanisms are commonly identified for shallow foundation systems and presented a case study on the developed a systematic approach for the seismic evaluation of building foundations. The proposed approach allows structural engineers to choose between an empirically based, relatively simple method, and a theoretical, more rigorous method involving numerical seismic response analysis, as may be appropriate for the seismic evaluation of foundations. Sharma, et al. (2014) presented the common practice to design the foundation is to provide, first the shallow foundation such as an isolated footing or raft to support the structure and if this is not adequate then deep foundation like piled foundation. It is discussed that, the choice between a raft and beam and slab raft foundation depends upon the soil properties and the weight of the building. Srilakshmi and Rekha (2011) carried out studies on Mat foundation. But there is lacking of theoretical studies on the mat foundation relatively. In addition to that, analysis of mat foundation by conventional methods viz., rigid method and flexible method are cumbersome and time consuming. The results of this analysis will be sustained to interest to designers of mat foundation.

Joy and Hassan (2014) studied the combined pile raft foundations provide an economical foundation option for circumstances where the raft foundation can satisfy the bearing capacity requirement, but fails to keep differential as well as maximum settlement below the maximum allowable limit. It had been established that augmenting features like the thickness of the raft, length of piles etch as decreased the settlement of the raft and on the other hand:decreasing 'spacing/depth' of piles has increased settlement of the raft. From the comparison of results, it has been found out that; installing high capacity piles at region with maximum load concentration and reinforcing the rest of the raft with medium capacity piles have the most important effect on significantly reducing maximum settlement and the differential settlement. Also, several resracer contributed on the deep foundation as one of the recommended choice in a weak strata. A cost analysis should be made to determine whether a mat or piles, in particular the type (steel, concrete, etc.), is more economical. In those cases where piles are used to control the settlement at marginal soil sites, care should be taken to utilize both the existing ground and the piles in parallel so that a minimum number are required. Robertson and Campanella (1983) also by Murad and Hani Titi (2004) carried out a study on Direct Cone Penetration Test methods for predicting the Ultimate Capacity for Friction Driven Piles. Their study presented an assessment of eight direct CPT methods to estimate the ultimate load capacity of square PPC friction piles driven into soils and The measured ultimate load capacity for each pile was determined from the load settlement curve using the Butler-Hoy method (1977). The ultimate load capacity of each pile was also determined using the CPT methods, and the static α and β methods. Their study showed that the ultimate pile capacity can be determined using the CPT data with the acceptable accuracy. The researchers concluded that the performance of the CPT methods may vary according to the performance used to determine the ultimate pile load capacity of the load test. Butler-Hoy method (1977) and Andras Mahler (2004) worked on four pile capacity methods were evaluated using CPT data for thirteen full scale axial pile load tests. To decide whether it is caused by the differing soil conditions or by differences in the pile installation process requires further studies. Nevertheless is clear that those methods represent a useful tool in Geotechnical design Raft and pile are two independent foundation systems, which are used extensively to support a variety of civil engineering structures. It is a common practice that the pile groups are provided by the pile cap (or raft). The distinction between the pile groups with and without a ground contacting cap has been brought out by a few researchers (Poulos 1968; Butterfield and Banerjee 1971; Davis and Poulos 1972) while analyzing pile group behavior. Zeavert (1957) was one of the earliest to prepare the concept of using deep foundation elements, particularly piles to reduce the raft settlement. Numerous researchers (Poulos 1994; Padfield and Sharrock 1983; Randolph 1983; Poulos et al. 1997) have studied the behavior of piled raft and evolved design procedures for the piles raft system.

Considering the above literature, it is observed that although the working and the efficiency of deep and shallow foundation are studied in several attempts, but isolated rigid pile foundation is not studied so far. There are severall attempts where raft foundation is proved to better choice if the superstructure and foundation syatem is asymmetrical. But the both are symmetrical then using this foundation may lead to uneconomical. To understand this the present study is focused to numerate the response and economy of islated rigid pile foundation system. Also very few studies show the soil-foundation-structure interaction effect. Considering this as a gap in research, present study oriented to study the performance of the isolated rigid pile footing for a mid rise building in a soft clay where the pile or raft foundation is a recommended choice. In this attempt the new foundation type, suitable for weak soil condition is proposed and its efficiency is evaluated by comparing the displacement and stress in the structure and foundation system for different weak soil conditions.

3 Methodology

In this study the 10 storey RCC building with 2 bays in X and Y directions with 3.0 m each storey height is considered. The building is modeled with FE program developed by the author as a part of her doctoral thesis, which was well tested with several test cases (Pallavi 2016). In this study the 2 numerical models are developed viz building with piled raft foundation system and the building with a rigid pile with isolated footing for a same soil condition. In this study the minimum thickness of isolated footing is considered to satisfy the punching shear criteria and the rigid pile with L/D ratio 20. For these two models the superstructre responses i.e. displacement and settlement are compared and suitability of proposed foundation i.e. Rigid piled isoated footing is evaluated. The foundation settlement and soil settlement are compared for a common points are studied to understand the efficiency of the proposed footing. The soil chosen in athe analysi is weak in nature and values of bearing capacity is taken from the following Table 2.

Table 2. Bearing capacity for different soil condition

Category	Types of rocks and soils	Presumed bearing value
Non-cohesive soils	Dense gravel or dense sand and gravel	>600 kN/m^2
	Medium dense gravel, or medium dense sand and gravel	<200 to 600 kN/m^2
	Loose gravel, or loose sand and gravel	<200 kN/m^2
	Compact sand	>300 kN/m^2
	Medium dense sand	100 to 300 kN/m^2
	Loose sand	<100 kN/m^2 *depends on degree of looseness*
Cohesive soils	Very stiff bolder clays & hard clays	300 to 600 kN/m^2
	Stiff clays	150 to 300 kN/m^2
	Firm clay	75 to 150 kN/m^2
	Soft clays and silts	<75 kN/m^2
	Very soft clay	Not applicable
Peat		Not applicable
Made ground		Not applicable

The preliminary design purposes, BS 8004 give presumed bearing values which are the pressures which would normally result in an adequate factor of safety against shear failure for particular soil types, but without consideration of settlement.

4 Numerical Modeling

In this study, the finite element program has been developed in C++ to understand the numerical formulation of each aspect of SSI modeling, including element formulation, interface implementation, soil nonlinear solvers and connections between soil, foundation and superstructure. The 10 storey square building has been modeled by the program developed in this study and the building, views have been created by sending the output file to the LS-PP freeware tool. The 30.0 m superstructure components, including beams and column have been modeled with 2 noded 3-D beam elements. The joints between beam and column are considered to be rigid. The connection between the foundation (raft/isolated slab) and first storey column is modeled as the rigid connections. The half space of size 20 × 20 × 20 m is modeled using as weak soil as a local soil domain which has a bearing capacity in the range and the engineering properties of the soil domain has been explained in detail in Table 3. The nonlinear behavior of the supporting soil is captured using associative Drucker–Prager material model. The meshing of the finite element model has been created by using GSA 2-D mesher. The SSI effect has been incorporated in the analysis by modeling the interfaces between soil and foundation and viscous boundary soil mass considered. The earthquake boundaries are model at the bottom of the soil block by modeling the bedrock and surrounding soil mass as a viscous boundary. The finite element model of SSI system has been developed to understand the coupled response of the soil and the structure for above mentioned foundation systems i.e. piled raft and rigid piled isolated footing.

Table 3. Details of the analysis and model details considered for the present analysis

Structural details for linear analysis						
Bay Width and height	Cross section of floor beam and columns	Ht. of building	R.C.C Grade Frame and rebar	Young's modulus (E)	Concrete density	Asymmetry shape
3.0 m	0.23 * 0.23 m	30.0 m	M20, Fe415	2.0×10^7 kN/m^2	2400 kg/m^3	Square
Foundation details for nonlinear static analysis						
Foundation type	pile L (m), Dia (m)	Raft/isolated slab	Poisson's ratio	E (MPa)	Vs (m/s)t	Density (kg/m^3)
Piled raft	9.0, 0.45	0.25 m	0.15	2.0×10^7	1200	2400
Isolated rigid pile	4.0, 0.25	0.5 m	0.15	2.0×10^7	1200	2400
Soil details for nonlinear static analysis						
soil type	Density (kg/m^3)	Friction angle, (°)	Poisson's ratio	E (MPa)	Material model parameters friction angle (δ) = 1/3 ϕ', friction coefficient $\mu = \tan\delta$	
Silt	1600	27	0.35	50	9°, 0.16	
Clay	1900	C = 15 kPa	0.30	75	0°, 0	
Silty clay	1700	20	0.32	63	12°, 0.12	

The modeling of the DSSI system for 10 storey building with piled raft layout has been modeled by using the engineering properties of the various modeling parameters of superstructure, soil, piles and the interface/contact explained in the Table 2. The configuration of raft and pile chosen as per the design guideline given by Is method. The reactions generated at the base of the column of the building have been observed and with that the initial set of pile dimensions has been worked out manually. As per the clause 6.6.2 in IS 2911 (Part 1 s.1, Rev. 2010), it has been recommended that for cast in situ friction piles the minimum spacing should be more than the 3 times the diameter of the pile shaft to ensure the zones of the soils in between the piles. Considering this pile foundation circular piles with 0.45 m and 9.0 m length have been modeled with the 3-D brick elements. The L-shape layout of piles accommodates the 23 piles spaced at 1.5 m c/c. The joints between the raft and pile have been modeled with the rigid connections. As per the IS 2950 (Part 1 s.43, Rev. 2013), The 0.25 m thick raft with the 0.5 m offset from all the sides of the base of the superstructure have been modeled with the 3-D brick elements. The raft thickness takes a minimum as per the piled raft design gs as the rigid platform for pile raft connection in numerical modeling.

For isolated piled raft footing the dimensions of footing decided on the punching shear criteria with considering the constant thickness of 0.5 m with 0.25 m offset from all sides of column i.e. 1.0 * 1.0 m footing dimensions and rigid piles of 0.2 m diameter and 4 m length (L/D ratio approximately 20) are placed at a 3d c/c distance with 0.1 m face offset from all sides. With that the 2*2 pile group can be accommodated with isolated footing slab (Figs. 1 and 2).

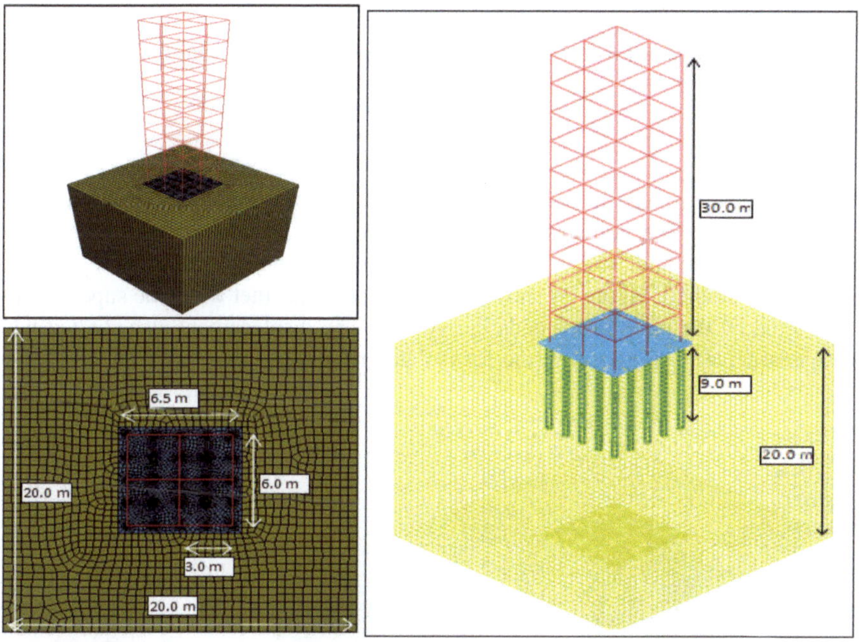

Fig. 1. Numerical model of piled raft foundation system with soil domain

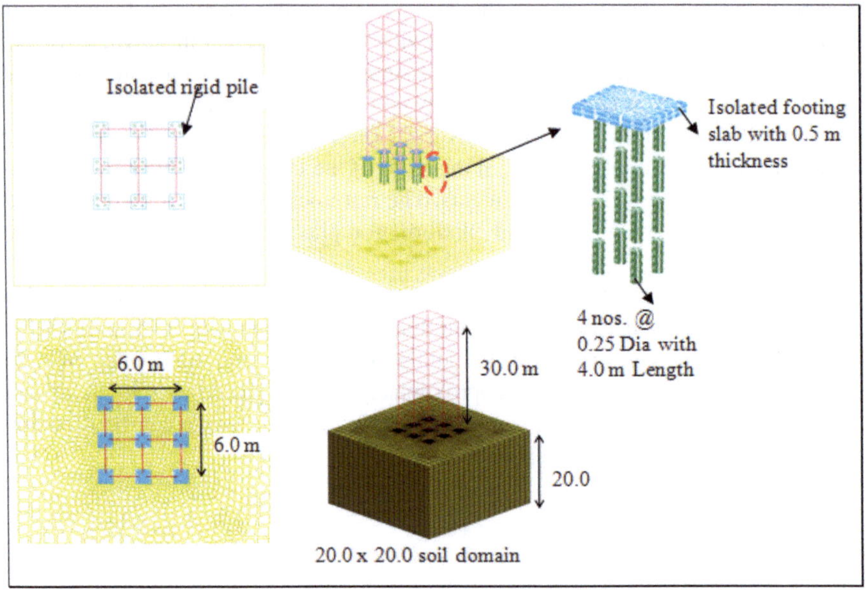

Fig. 2. Numerical model of isolated footing with rigid piles with soil domain

5 Results

The exact impact change in foundation syatem and type of soil has been studied for system responses. The nolinear analysis is carried out for the gavity load condition including the slab and wall weight applying at the nodes of each storey and the responses are studied. Figure 3a shows the system displacemet when the superstructure is supported by the piled raft foundation system. The displacement also studied when the superstructure is founded on the for isolated rigid piled foundation (Fig. 3b).

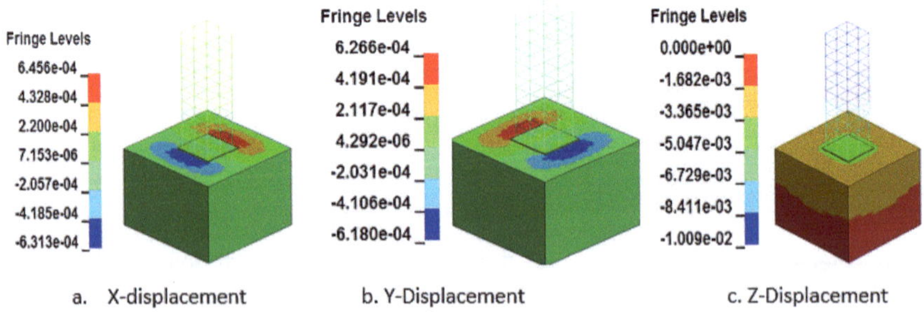

Fig. 3(a). Displacement in SSI system for structure with isolated rigid pile foundation

From Fig. 3a and b it has been observed that the lateral direction displacemnts are negligible but settlement of the system is predominant.In lateral direction near foundation periferi experinces a higher displacemnts that the inner part, althougt the variation is small in respective directions. For settlement the enite foundation is expencing the larger displacement values and in that also piles experiencing the higermovemnt in vertical direction. In Isolated rigid pile foundation the bottom and to p of the rigid piles shows the same settkemnt as the isolated slab but middle part of the pile settles more than the top and bottom. In piled raft foundation raft slab shows very less settlement than the connected piles. Also both pile ends shows the same settlement but more at the middle part (Fig. 3c).

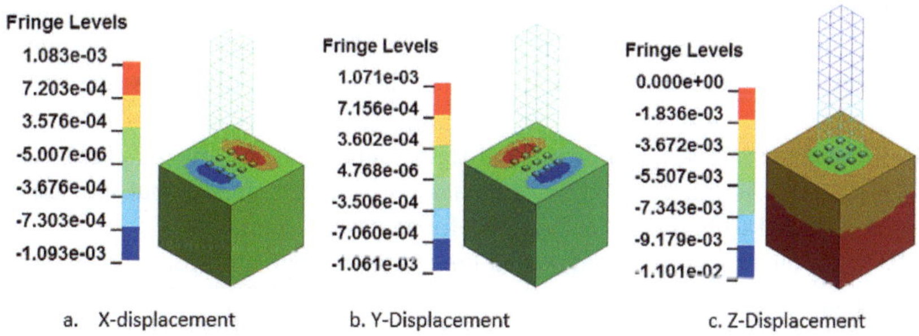

Fig. 3(b). Displacement in SSI system for structure with isolated rigid pile foundation

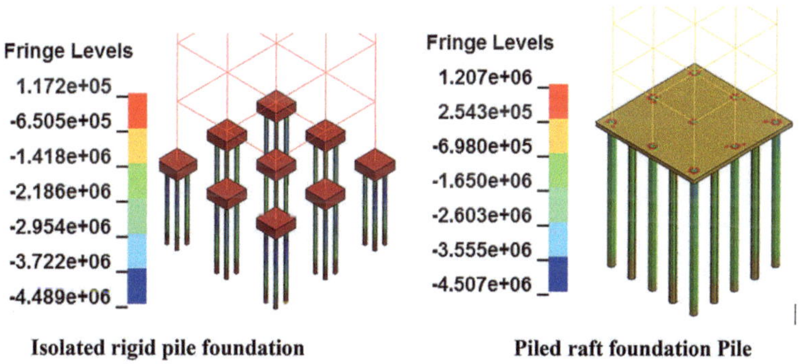

Fig. 3(c). Pile settlement

Comparing the peak settlement in foundations for isolated rigid pile and piled raft the overall peak settlemt in rigid pile system is about 6% lesser than the piled raft system.

The displacemnts are compared for both the foundation types (Fig. 4a). Apart from displacement the system peak stresses also observed for both the foundation system (Fig. 4b).

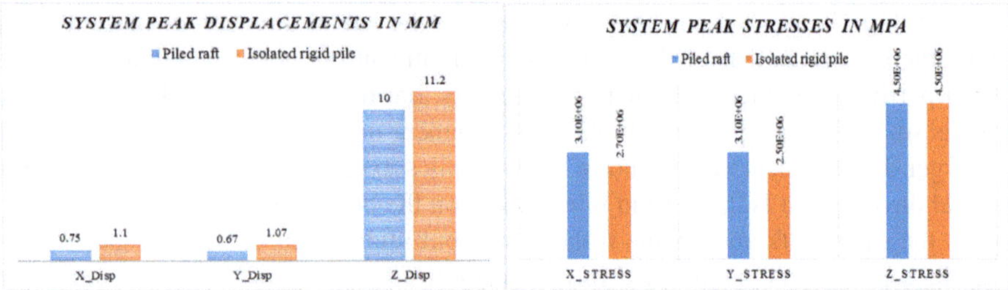

4a. 4b.

Fig. 4. Comparative response for both foundation system

With comparison of the peak displaements for both foundation system including piled raft and Isolated rigid pile, it is observed that isolated rigid type foundation experiences 12% more displacement. Lateral displacement also vary in with same trend.

But found the stresses less in isolated rigid foundation less by 14% than the piled raft. This various is found to be valid for lateral direction stresses but the stress in z direction is found to be same for both the foundation system.

The study also focused to understand the effect on responses (for different soil type in a weak category (Fig. 5).

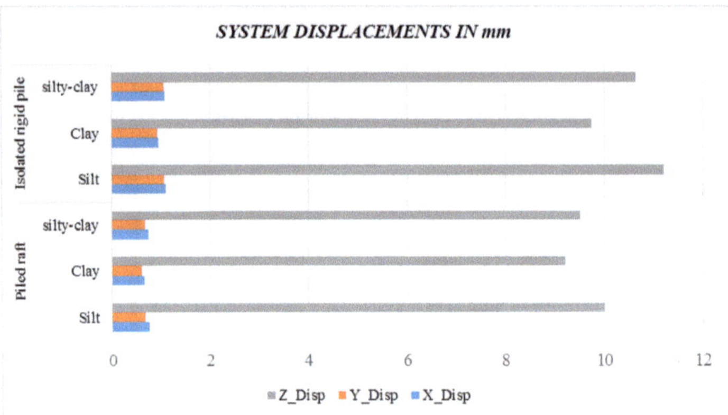

Fig. 5. Comparative response for both foundation system for different soil types

To analyse this the soil chosen includes the Silty clay, clay and silt and engineering properties are shown in Table 1.

It is found that the in piled raft foundation system, clay type of soil is better to restrai the settlement than silty caly and silt is the one which offered larger settlement in comparison wth other soil. In this study the difference in largest to smallest settlement value is foun to be around 9%. And for lateral displacement this difference is found to be very negligible as 3%. And the same observation is noted for Isolated rigid pile footing with the difference in larger and smaller value of settlement is 14%, and the difference is negligible for lateral displacements.

Always soil structure interaction is numerically expensive analysis and thus to understand the numerical model with that fascets the FE model is studied. Table 4 shows the numerical attributes of each model to understanfd the numerical cost of each model.

Table 4. Numerical statistics

Foundation system	No. of elements	DoFs	Least element size in FE model (m)	Concrete used for foundation system (Kg)
Piled raft	59457	176574	0.1	11.11×10^4
Isolated rigid pile	78695	285217	0.03	9.7×10^4

It has been observed that the piled raft model has less degrees of freedom which is the measure of size of stiffness matrix and thus commands the numerical time. Also considering the same meshing type the number of elemnts in piled raft model is lesser than the Isolated rigid pile model.

To understand the economical aspect of the proposed foundation system the volume of the concrete required is estimated for both the foundation system and it is found that the islated pile raft foundation requries less conctrete (about 14% less) for construction than the piled raft foundation, thus proves to be economical over piled raft foundation for same load and superstructure and soil condition.

6 Conclusions

The following are the conclusions drawn from the present study.

6.1 Displacements and Stresses

The displacements and the stresses are found to nearly same for isolated rigid pile foundation system and piled raft foundation system. In these camparion although the peak system displacement for isolated rigid pile foundation system vary from piled raft system but the streese are same. The piles settles less in case of isolated rigid pile system than the piled raft system. With this the proposed foundation system is found to be equally good and reccommanded choice for weak soil for the static load conditions.

6.2 Effect of Different Soil Type

It is infer that the soil types varies the sytem responses and found that the C-soils are better soil type for less settlement and lateral displaccment contrary to that phi soil offers more settlement to the foundation system. With this the propsed foundation system is compatable withpiled raft with negligible deviation of the responses. So for stastic loads the isolated rigid pile foundation is acceptable foundation type for all types of soil with considerable response deviation.

6.3 Numerical Statistics and Economical Aspect

It has been inferred from numerical statitics that the isolated rigid pile foundation is creates numerically more expensive than the piled raft model. Also, proposed foundation system, is economical over a piled raft foundation system.

References

Baidya, D.K.: Earthquake resistant design of shallow foundation. In: Chapter 8 GT 102 NPCBEERM, MHA (DM), pp. 1–5. https://doi.org/10.1007/1-4020-2528-9_11

Ghazavi, M., Mahali, A.S.: Determination of seismic bearing capacity of shallow strip footings on slopes. In: Proceedings in the 8th Symposium on Advances in Science and Technology (8th SASTech), Mashhad, Iran, pp. 1–10 (2014). 8thSASTech.khi.ac.ir https://doi.org/10.1061/(asce)1532-3641(2006)6:3(176)

Joy, N.J., Hassan, H.: Study on settlement characteristics of combined pile raft foundation founded on sand with various arrangements of piles using plaxis – 3 D. International Journal of Emerging Technology and Advanced Engineering **4**(10), 1–10 (2014). https://doi.org/10.1179/1939787914Y.0000000052. ISSN 2250-2459, ISO 9001:2008 Certified Journal

Kaneko, O., Nakai, S., Futaki, M., Arai, H.: Evaluation system of seismic capacity for building foundations and case studies on buildings in Sendai City. In: International Symposium on Engineering March 1–4, Tokyo, Japan, pp. 1200–1208 (2012)

Abu-Farsakh, M.Y., Titi, H.H.: Assessment of direct cone penetration test methods for predicting the ultimate capacity of friction driven piles. J. Geotech. Geoenvironmental Eng. **130**(9) (2004). https://doi.org/10.1061/(ASCE)1090-0241(2004)130:9(935)

Nirmal, J.J., Hashifa, H.: Study on settlement charecteristics of combined pile raft foundation founded on sands with various arrangements of piles using plaxis 3D. Int. J. Emerg. Technol. Adv. Eng. **4** (2014). https://doi.org/10.1088/1757-899x/342/1/012002

Polous, H.G.: The piled raft foundation for the Burj Dubai—design and performance. IGS—Ferroco, Terzaghi, Oration (2008)

Poulos, H.G.: Settlement prediction for bored pile groups, deep foundations on bored and auger piles. Balkema, Rotterdam (1993). https://doi.org/10.1061/40865(197)1

Poulos, H.G., Davis, E.H.: Pile Foundation Analysis and Design. Wiley, New York (1980). https://doi.org/10.1061/40865(197)1

Pallavi, B., Neelima, D.S.: Seismic soil structure interaction analysis of piled-raft supported asymmetrical. Buildings, Ph.D. thesis (2016). shodhganga.inflibnet.ac.in/handle/10603/136238

Frank, R.: Some aspects of soil-structure interaction according to Eurocode 7 geotechnical design, CERMES (Soil Mechanics Teaching and Research Centre, ENPC-LCPC) (2006). https://doi.org/10.1080/17499510902788843

Rajashekhar Swamy, H.M., Krishnamoorthy, P.D.L., Bhavikatti, S.S.: Relevance of interface elements in soil structure interaction analysis of three dimensional and multiscale structure on raft foundation. Electron. J. Geotech. Eng. (2011). 10.1.1.640.3963&rep=rep1

Randolph, M.F.: Design methods for pile groups and piled rafts. In: Thirteenth International Conference on Soil Mechanics and Foundation Engineering, India (1994)

Roy, D.: Design of shallow and deep foundations for earthquakes. J. Geotech. Earthq. Eng., 1–8 (2013). https://doi.org/10.13140/rg.2.1.1492.6889

Raj, D., Bharathi, M.: Analysis of shallow foundation on slope: a comparative study. In: International Symposium Geo hazards: Science, Engineering and Management, Kathmandu, Nepal, Paper No. LF-16 (2014). https://doi.org/10.1080/19386362.2017.1416971

Srilakshmi, G., Rekha, B.: Analysis of MAT foundationusing finite element method. Int. J. Earth Sci. Eng. **04**(06 SPL), 113–115 (2011). ISSN 0974-5904, https://doi.org/10.1061/9780784412688.056

Sahoo, R.R.: The behavior of eccentrically loaded shallow foundations on granular soil. A thesis submitted in partial fulfillment of the requirements for the degree master of technology, IIT rurkee (2013). https://doi.org/10.1016/j.compstruc.2017.07.018

Sharma, S.M., Vanza, M.G., Mehta, D.D.: Comparison of raft foundation and beam & slab raft foundation for high rise building. International Journal of Engineering Development and Research **2**(1), 572–75 (2014). ISSN: 2321-9939, https://doi.org/10.1109/icetce.2012.35

Lateral Resistance of Steel Pipe Pile with Wings by Static Cyclic Loading Tests

Toshio Kobayashi[1,2](✉) and Yuji Miyamoto[1]

[1] Department of Architectural Engineering, Division of Architecture, Osaka University, Osaka, Japan
kobayashi@sansei-inc.co.jp
[2] Sansei Inc, Tokyo, Japan

Abstract. Steel pipe piles subjected to lateral loads exhibit toughness even after yielding of the steel, and thus possess a high earthquake resistance. A steel pipe pile with wings attached to its tip can be driven into the ground by rotation. However, because the wings are larger than the pile diameter, the soil around the pile shaft is loosened, resulting in an alteration of the lateral subgrade reaction of the pile. In the present study, static lateral loading tests were conducted to ascertain how the lateral resistance of a pile was affected by the attachment of wings and the wing diameter, and to examine the occurrence of secular change after one year.

The soil deposit at the test site was a weak volcanic ash. Four types of piles of diameter (*Dp*) 267.4 mm and length 5000 mm were tested, namely, piles without wing, piles with wing diameters 2.3*Dp* and 3.0*Dp*, respectively (used to investigate the effect of the wing diameter), and a pile with wing diameter 2.3*Dp* after one year of its insertion into the ground (used to evaluate the recovery of the loosened soil). The cyclic load applied to a pile was gradually increased by displacement control at the pile head, and the lateral displacement of the pile head and the bending moment along the pile were measured. The test results showed that (1) the lateral resistances of the winged piles were smaller than that of the piles without wing, owing to the loosening of the surrounding soil by the wings; (2) the maximum bending moments of the winged piles tended to be smaller than that of the piles without wing for a given pile head displacement; (3) the pile wing diameter had little effect on the lateral resistance; and (4) there was no aging in the lateral resistance of the loosened soil around a winged pile, even one year after insertion of the pile.

1 Introduction

Many versions of a steel pipe pile with a steel spiral or flat wing welded to the tip or peripheral surface of the pile (hereafter, winged pile) have been developed and put to practical use [1]. Construction projects that utilize winged piles generally involve the driving of a pile shaft (hereafter, pile) into the ground through the application of a rotational torque. The process is hereafter referred to as rotary installation.

Compared to the use of cast-in-place concrete piles and embedded piles, winged piles enable reduction of the construction byproducts by decreasing the generation of waste soil. In addition, the employed construction equipment is smaller and quieter and

D. Choudhury et al. (Eds.): GeoMEast 2018, SUCI, pp. 40–52, 2019.
https://doi.org/10.1007/978-3-030-01920-4_4

produces less vibration, thus facilitating construction within limited spaces and reducing the environmental load of the piling process. Piles with large wings have become come in recent times, with their applications expanding to large buildings [2]. However, the diameter or equivalent circular diameter of a winged pile (the wing diameter *Dw*) is larger than the diameter of the pile (pile diameter *Dp*). The rotary installation of a winged pile thus disturbs the surrounding soil of the pile. Arai et al. [3]. investigated the disturbance through experiments using a model soil layer, and found that the disturbance varied with the pile driving rate for a given rotational velocity. Kuze et al. [4, 5] and Mori et al. [6]. investigated the lateral resistance of winged piles and collected soil quality data from the area through which the wings passed during the pile installation. They also conducted soil tests on the pre- and post-installation grounds to ascertain the occurrence of changes in the soil, such as regarding the density, structural strength, compressive strength, and rigidity. In addition, a lateral loading test was conducted to investigate the ground reaction force and the bending moment generated in the pile body.

However, in all these previous studies, the loading tests were conducted using a unidirectional static loading test. The effects of repeated loading, as occurs during an earthquake, could thus not be established. Further, the literature [7] recommends a certain number of days for curing cohesive and sandy soils because sufficient curing is required to restore the strength of the ground after disturbance by the pile installation. On the contrary, to the best knowledge of the present authors, there has been no research in which cyclic lateral loading experiments were performed on winged piles, taking into consideration the recovery of the ground strength or temporal changes in the ground. In the present study, a pile with two flat plate wings was subjected to cyclic lateral loading about its original position to examine the following:

(1) Effect of the pile wings on the lateral resistance
(2) Effect of the wing diameter ratio on the lateral resistance
(3) Occurrence of aging of the lateral resistance after installation of the pile.

2 Lateral Loading Experiment

2.1 Experiment Site

The experiment was conducted in Omitama City, Ibaraki Prefecture. Figure 1 shows the *N*-value distribution, the soil column, and the main soil test results for the ground of the test site. The soil between the ground surface and 0.55 m below ground level (GL) consists of andosol, that between GL-0.55 m and GL-2.2 m consists of weathered volcanic ash, that between GL-2.2 m and GL-3.2 m consists of volcanic ash, and that beyond GL-3.2 m consists of sand.

2.2 Experiment Pile

The piles used to perform the present experiment were fabricated from JIS G 3444 general structural carbon steel (STK) pipes. Each pipe had an outer diameter φ of

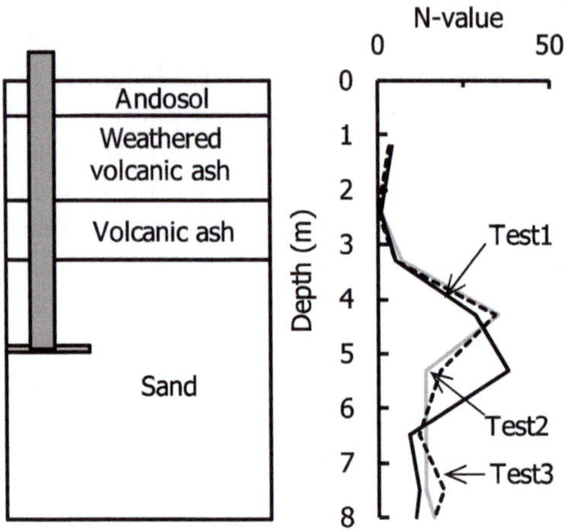

ρ (t/m³)	c (kN/m²)	φ (°)	Vs (km/s)	ν
1.28	29.3	16.9	0.13	0.29
1.34	35.1	10.6	0.10※	0.44※
1.64	89.6	9.3	0.16	0.47
1.80			0.15※	0.47※

Fig. 1. Test results of the soil at the experiment site

267.4 mm and material thickness t of 9.3 mm. The steel was of grade 490. The yield bending moment M_y of the test pile was 148.1 kN·m. The winged pile was fabricated by welding two flat plates to the tip of the steel pipe. The edge was prepared such that the inclination angle of each wing was 10° (Fig. 2). A triangular excavation blade of a size that fitted into the inner diameter of the steel pipe was attached by welding to the tip of the winged pile (hereafter, ST) (Fig. 3). The tips of all the test piles were closed, and all had a shaft length of 5 m. To measure the bending strain on the inner surface of a pile shank during the experiment, strain gauges were attached to 50 points dispersed at 100 mm intervals in the loading direction.

Fig. 2. Test pile with wing (W30)

Fig. 3. Test pile with piles without wing (ST)

Table 1 lists the experimental cases. A total of four test piles were fabricated, including one ST, which was used as the reference test pile. There were also two winged piles with wing diameter ratio *Dw*/*Dp* values of 2.3 and 3.0, tagged "W23" and "W30," respectively. The regions of the soil through which the wings of these two piles traversed during their installation were respectively of diameters 606 and 800 mm, which are the diameters of the circles traced by the respective wings.

Table 1. Test cases

Test number	Specimen		Completion date of pile installation	Start of lateral loading test	Days between pile installation and lateral loading test
	Full name	Abbreviation			
1	Screw pile without wings	ST	18 Oct 2016	28 Dec 2016	71
2	Screw pile with wings (wing diameter ratio 2.3)	W23	17 Oct 2016	10 Dec 2016	54
3	Screw pile with wings (wing diameter ratio 3.0)	W30	17 Oct 2016	23 Jan 2017	98
4	Screw pile with wings (wing diameter ratio 2.3) at 1 year after installation	W23A	17 Oct 2016	24 Nov 2017	403

Table 1 gives the number of days required for the installation of each test pile, and the number of curing days until commencement of the experiment. The test pile and reaction pile were placed by rotary installation, with the pile head struck to at GL + 200 mm above ground level.

2.3 Loading Method

Figure 4 presents an overview of the experimental apparatus. The loading system consisted of the test pile, a reaction pile, and a hydraulic jack. A base plate of thickness 25 mm was welded to the pile head of the test pile and reaction pile for attachment of a load beam and other components. Four STK490 steel pipes with diameter φ of 406.4 mm and thickness t of 12.7 mm were used to fabricate the reaction pile. To achieve the reaction force, the heads of the four reaction pile pipes were connected in a square configuration using an H-shaped steel connector.

The hydraulic jack was attached between the load beam and the reaction block, which was joined by high-strength bolts to the test pile. The experiment was conducted under the condition of free rotation of the head of the test pile. The loading height was GL + 400 mm. In addition to the pile strain, the lateral displacements and out-of-plane displacements at the loading height of the test pile and reaction system were measured

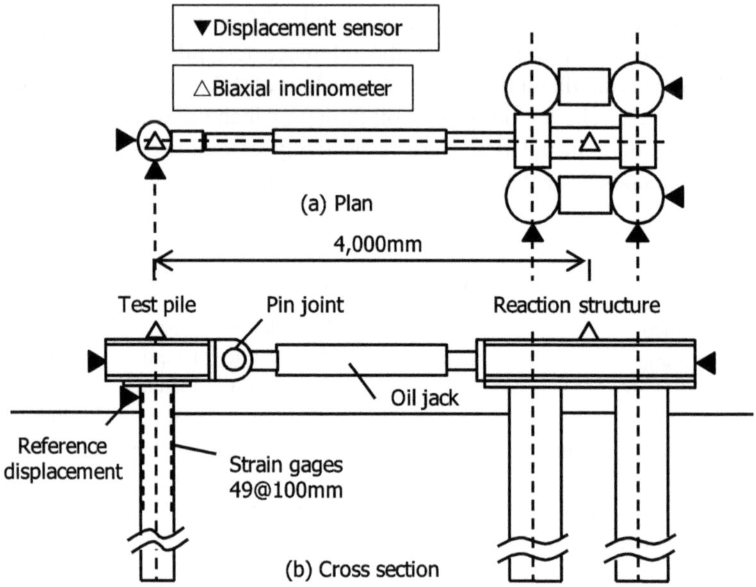

Fig. 4. Experimental apparatus and sensor locations

using a displacement sensor installed at the center of the pile head. Inclinometers installed at the center of the head of the test pile and on the reaction system was used to measure the tilt angle. The lateral displacement of the test pile was measured at two points, namely, the loading height and GL + 100 mm. The lateral displacement at GL + 100 mm was adopted as the reference displacement. Figure 5 shows the diagram of the load cycle during the experiment. The experiment controlled was made by the reference displacement.

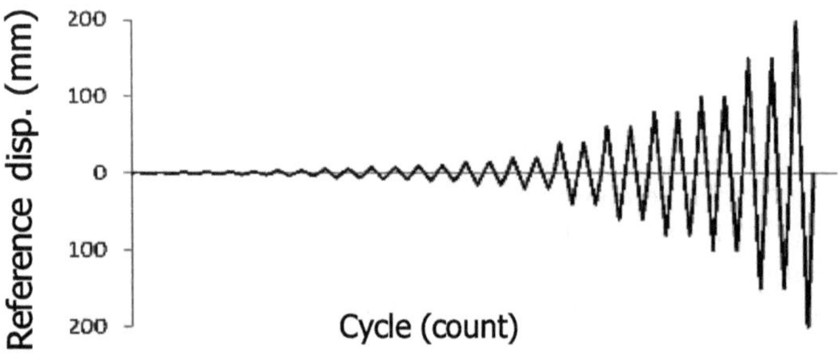

Fig. 5. Cyclic conditions of the reference displacement

3 Lateral Loading Experiment

The experimental results for the ST and W23 piles are presented and discussed in this section, with a focus on reference displacement amplitudes of up to 20 mm, which is about 10% of the pile diameter, and displacement amplitudes of up to 60 mm, at which the ST pile does not undergo plastic deformation.

3.1 Pile Head Load Versus Reference Displacement

Figure 6 shows the relationship between the pile head load and the reference displacement. The initial loading curve for W23 reveals a decrease in rigidity from the beginning of the loading. The difference between the hysteresis properties in the presence and absence of the wings is attributed to the differing rigidities of the respective piles, with the rigidity of ST being higher than that of W23. Hence the plastic deformation of the ground around ST is relatively small.

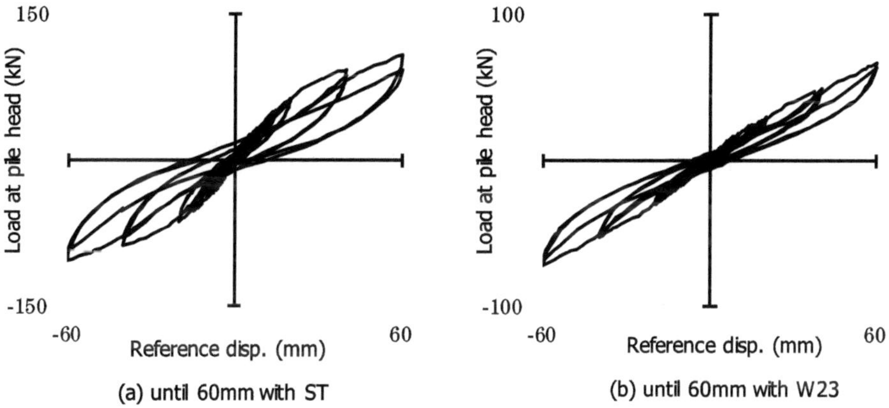

Fig. 6. Pile head load versus reference displacement

3.2 Ground Deformation

Figure 7 shows the ground deformation. The figure shows the cracked condition of the ground surface at the peak load during the first cycle for reference displacement amplitudes of 20 and 60 mm. In the case of a reference displacement amplitude of 20 mm, many cracks occur around ST and the crack growth is remarkable, whereas there are almost no cracks around W23, with an opening formed between the pile and the surrounding ground. In the case of a reference displacement amplitude of 60 mm, there is increased cracking around both ST and W23, and the crack growth is significant. The crack propagation is especially remarkable for ST, and there is increased cracking in the direction perpendicular to the loading axis, in addition to cracking in the direction of the loading axis. The crack distribution tended to be symmetrical over the entire observation surface, with the crack propagation having a "wedge"-like shape spreading at about 45° on both sides of the loading axis.

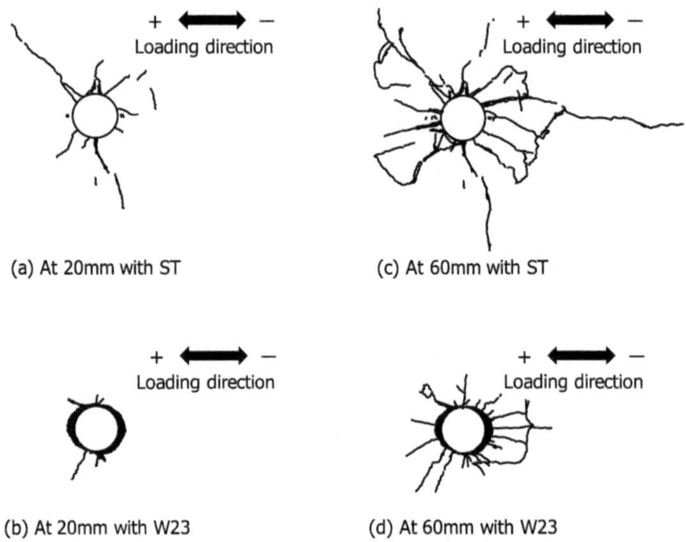

Fig. 7. Cracking of the ground surface around the test piles

3.3 Properties of Bending Moment Distribution

Figure 8 shows the bending moment distribution in the depth direction for reference displacements of 20 and 60 mm. The bending moment was calculated based on the readings of the strain gauges attached to the pile. The bending moment distribution at the same reference displacement was compared. The bending moment generated in W23 became smaller than ST and the maximum bending moment depth of W23 tended to be deep.

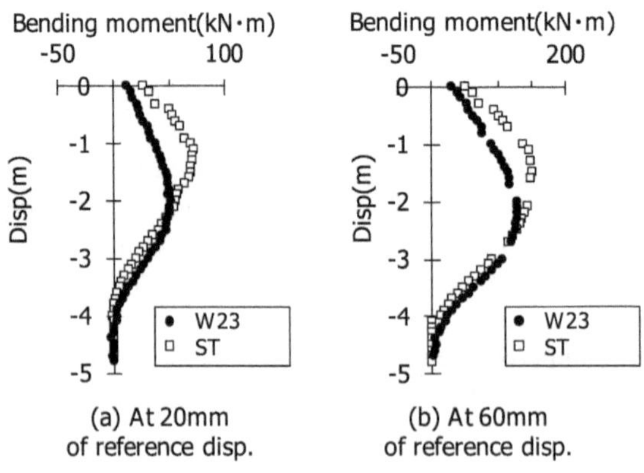

Fig. 8. Comparison of the bending moment distributions of W23 and ST

3.4 Subgrade Reaction of Pile Displacement at Ground Surface Position

The subgrade reaction and displacement distributions were determined from the bending moment of the pile obtained from the attached strain gauges and the curvature factor distribution [8]. In the case of the subgrade reaction distribution, the bending moment distribution of the pile in the depth direction for a given pile head displacement was first approximated using a sixth-order polynomial and the least squares method. The first derivative of the bending moment distribution approximated by the function was then used to calculate the shear force, and the second derivative was used to calculate the subgrade reaction. These were based on the imposed conditions of the subgrade reaction at the ground level being 0 and the pile head shear force equal to the pile head load, with the pile tip position (GL − 4.8 m) set as the boundary, where the bending moment and subgrade reaction are 0. The foregoing is expressed by Eq. (1).

$$\mathrm{M(z)} = a_0 + \sum_{i=1}^{6} a_i z^i \tag{1}$$

Here, $a_0 = M(0) = H \cdot ht, a_1 = H, a_2 = 0, a_3 - a_6$ are arbitrary parameters, H is the pile head load, and ht is the pile protrusion height. The distribution of the pile curvature for an arbitrary pile head displacement can be approximated by the pile displacement using a fifth-order polynomial, provided that the condition of zero curvature at the pile tip is fulfilled. The approximate function can thus be expressed with respect to the curvature as in Eq. (2).

$$\emptyset(z) = b_0 + \sum_{i=1}^{5} b_i z^i \tag{2}$$

Here, b_0–b_5 are arbitrary parameters.

The displacement is calculated by performing a second-order integration of the approximate expression in Eq. 2. The condition that the reference displacement was the measured value was imposed on the function, along with zero displacement of the pile tip. The approximate expression of the displacement used in the present study is thus as in Eq. (3).

$$\mathrm{y(z)} = y_0 + \theta \cdot z + \sum_{i=0}^{5} \frac{1}{(i+1)(i+2)} b_i z^{(i+2)} \tag{3}$$

Figure 9 shows the relationships between the subgrade reaction and the pile displacement at GL − 0.5 m and GL − 1.5 m for reference displacements of 20 and 60 mm. In addition, Fig. 11 reveals a slip in the hysteresis of W23 compared to that of ST. The slip is particularly significant at the shallow depth of GL − 0.5 m, becoming quite small at a depth of GL − 1.5 m. However, no significant variation of the hysteresis properties of ST with depth was observed. This indicates a small effect of the weight of the soil on the restraint in the region affected by the ground surface, and that the slip in the hysteresis of W23 may be attributed to the significant effect of the pile wings on the surrounding ground.

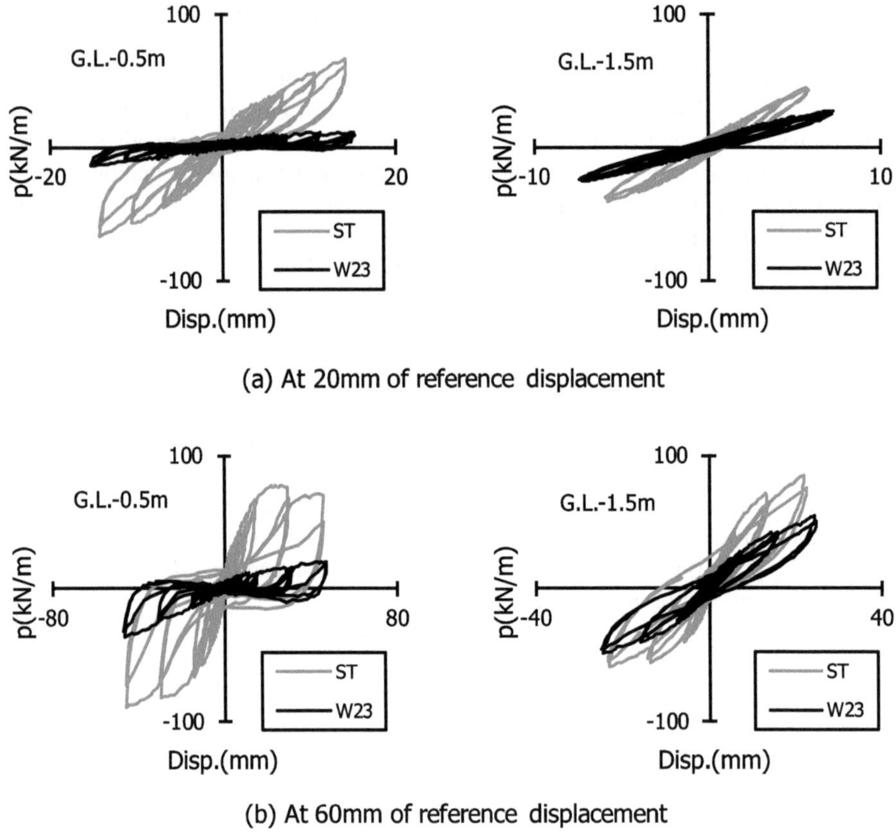

Fig. 9. Hysteresis curves: subgrade reaction versus pile displacement relationships for ST and W23

3.5 Effect of Pile Wings

The experimental results revealed that ST and W23 were characterized by differing hysteresis properties, ground deformations, bending moment distributions, and subgrade reactions. The hysteresis properties of W23 indicated that the ground around the pile underwent plastic deformation early during the loading, because the reference displacement revealed the occurrence of slip at 20 mm. Further, the observed ground reaction force in the depth direction indicated that the ground rigidity of W23 was less than that of ST, and the former thus provided less resistance when a lateral load acted on the pile, with the hysteresis exhibiting slip. It can therefore be inferred that the lateral resistance of a winged pile is decreased by its wings loosening the surrounding ground during the rotary installation process.

Figures 10 and 11 show the inside of the pile's surrounding ground of ST and W23 for a pile displacement of 200 mm. Both figure was taken from the gap generated on the behind of the pile opposite to the loading direction. Difference was found in the unevenness of the peripheral surface of the pile and the amount of soil adhering to the pile.

Fig. 10. Underground conditions of ST for a reference displacement of 200 mm

Fig. 11. Underground conditions of W23 for a reference displacement of 200 mm

3.6 Effect of Wing Diameter Ratio

Figure 12 shows the relationships between the pile head load and the reference displacement for W23 and W30, while Fig. 13 shows the bending moment distributions. It can be observed from these figures that the wing diameter ratio has no effect on either the hysteresis properties or the bending moment distribution. The same applies to the ground deformation, with no significant variation of the amount or progression of the cracking with the wing diameter ratio observed. We thus conclude that the wing diameter ratio factor of approximately 1.3 between the two winged piles had almost no effect on the lateral resistance.

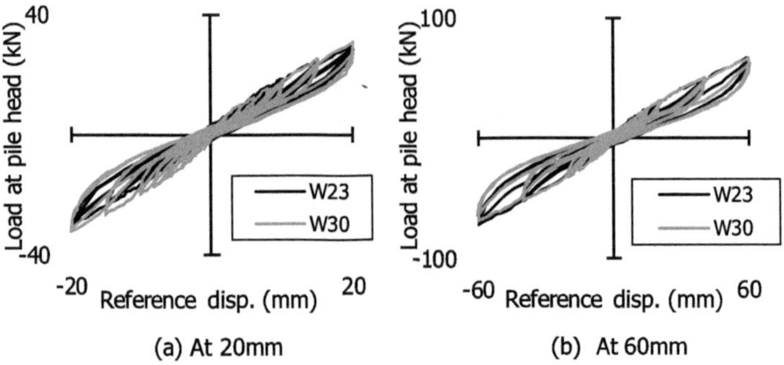

Fig. 12. Head load versus reference displacement relationships for piles with different wing sizes

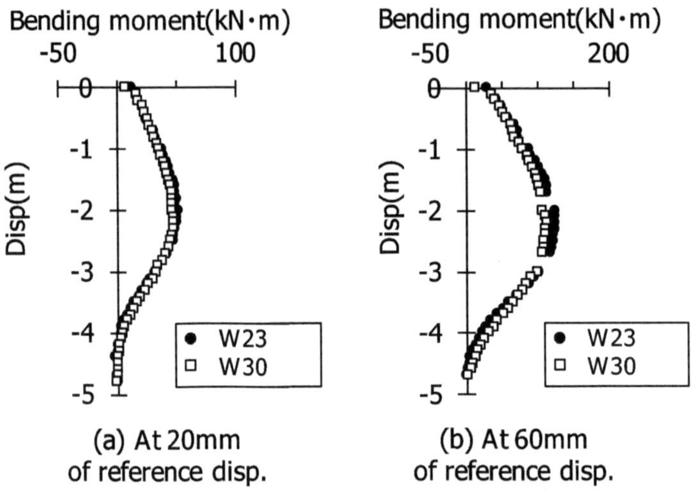

Fig. 13. Bending moment distributions for piles with different wing sizes

3.7 Aging of Lateral Resistance of Pile After Installation

Figure 14 shows the relationship between the pile head load and the reference displacement for W23 piles with different curing periods of up to 349 days. Figure 15 shows the bending moment distribution. As can be observed from the figures, the curing period has a small extent effect on the hysteresis properties or the bending moment distribution. The results of the present experiments thus indicate that, even after a curing period of over one year, there is a small extent enhancement of the lateral resistance of a winged pile placed by rotary installation.

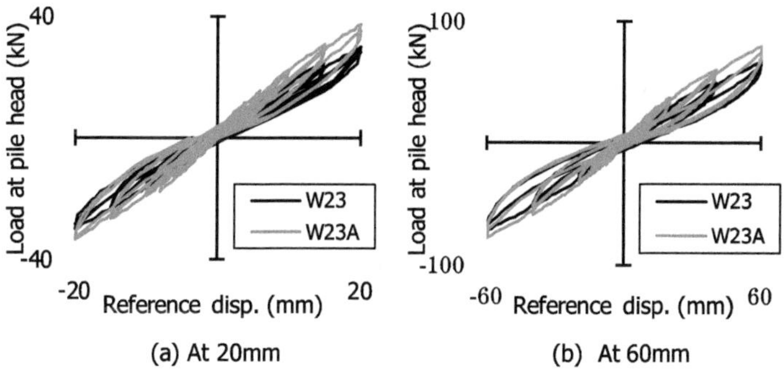

Fig. 14. Head load versus reference displacement one year after installation of the pile

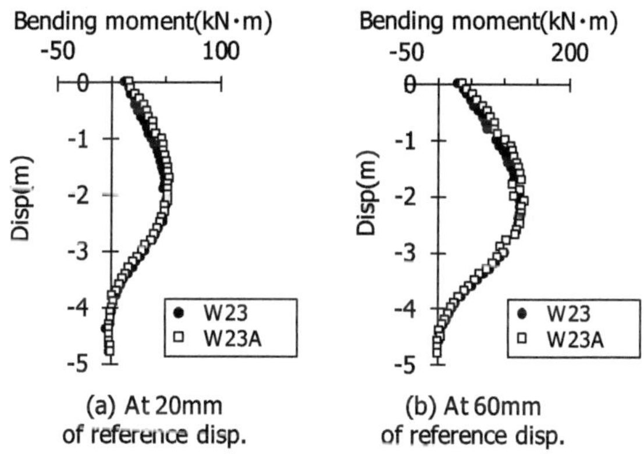

Fig. 15. Bending moment distribution one year after installation of the pile

4 Conclusions

In the present study, the lateral resistance of winged steel pipe piles was investigated through the conduction of static cyclic loading tests. The findings are as follows:

(1) The lateral resistance of a winged pile is smaller than that of piles without wing owing to the loosening of the surrounding soil in the former.
(2) The maximum bending moment of a winged pile tends to be smaller than that of piles without wing for a given pile head displacement.

(3) There is little difference between the lateral resistances of piles with equivalent diameters 2.3 and 3.0 times the pipe diameter.
(4) Even one year after construction, there is no aging of the lateral resistance of the loosened soil around a winged pile.

Acknowledgments. The authors would like to express their special gratitude to Dr. Eng. Kashiwa and Dr. Eng. Nakano for their useful suggestions for this study. The study was supported by JSPS KAKENHI (Grant No. JP17H03342).

References

1. Tsuchiya, T.: The trend of a screwed steel pile and the problems in the future. Found. Eng. Equip. Mon. **34**(11), 2–6 (2006). [in Japanese]
2. Inoue, N.: Recent trend of small-diameter rotating intrusive steel pipe pile construction method. Found. Eng. Equip. Mon. **41**(2), 36–39 (2013). [in Japanese]
3. Jun, A.M., Mamoru, F., Makoto, N., Gorou, K., Emi, K.: Effect of penetration velocity on turbulence of soil around the screw pile. In: Summaries of Technical Papers of Annual Meeting Hokkaido Chapter, Architectural Institute of Japan, pp. 685–686, August 2004. [in Japanese]
4. Naoya, K., Shoichi, N., Nmihiko, I., Tadashi, M., Yoshiaki, T., Chihiro, A.: Maintenance of technical standard about building site, soil and foundation (Part 6: Study on Lateral Loading Behaviour of a Steel Pipe Pile with Wings). In: Summaries of Technical Papers of Annual Meeting Kanto Chapter, Architectural Institute of Japan, pp. 407–408, August 2011. [in Japanese]
5. Kuze, N., Suemasa, N., Inoue, N., Futaki, M.: Property changes of soil around a rotary press-in pile with wings and lateral loading behavior of the pile. AIJ J. Struct. Constr. Eng. **81**(727), 1455–1465 (2016). [in Japanese]
6. Mori, G., Hayashi, M., Shinohara, T.: Lateral loading behavior of screw steel pipe pile. In: Japan National Conference on Geotechnical Engineering, vol. 35, No. 727, pp. 1755–1756, June 2000. [in Japanese]
7. Method and Commentaries for Lateral Load Test of Piles [in Japanese]. The Japanese Geotechnical Society, May 2010
8. Kashiwa, H., Kobayashi, T., Miyamoto, Y.: Evaluation of lateral resistance of steel pile with wings in cyclic loading tests. AIJ J. Technol. Des. **24**(56), 99–104 (2018). [in Japanese]

Reinforced Embankments on Soft Soil, Focus on Lateral Spreading

George S. Fahmy(✉)

DAR AL-HANDASAH Consultants (Shair & Partners), Cairo, Egypt
george.fahmy@dargroup.com

Abstract. Embankments on soft soil supported by basal reinforcement and pile-like elements are analyzed regarding the lateral spreading. A series of three-dimensional well- instrumented model tests at a scale of 1:3 has been carried out to investigate spreading and shear stresses at the base of a reinforced embankment and the results were verified by FE-model. Moreover, different parameter variations such as embankment slope, embankment height, number of reinforcement layers, and deformation modulus of the soft underground had been analyzed through a detailed parameter study on the prototype using 3D-FEM. A modified analytical method to calculate the spreading force in reinforcement had been derived considering these different parameters. Dimensionless factors are developed to include these parameters in determining the spreading forces.

1 Introduction

The construction of embankments on weak or soft soils such as peat is normally treated using a soil replacement method or by introducing pile-like-elements into the soft layer to partially support the embankment. In the slope zone of the embankment the underground is subjected to additional lateral stresses due to the spreading effect of the slope. The stability of the bearing system is influenced by spreading stresses which cause horizontal displacement of the pile-like-elements or the toe of the embankment slope.

Kempfert et al. (1997) pointed out that the horizontal forces must be transferred to reinforced elements, such as horizontally-placed geosynthetics reinforcement. Figure 1 shows the structural system and the load transfer mechanism of the lateral spreading.

The spreading stresses concentrate at the toe of the slope and lead to a horizontal displacement. Kempfert et al. (1997) concluded that these lateral forces must be transferred to geosynthetic reinforcement at the base of the embankment which can prevent the embankment from sliding. Figure 2 illustrates the mechanism of load transfer of the lateral forces to the reinforcement. The lateral earth pressure forces develop outward shear stresses and lateral displacement at the base. The reinforcement must sustain the shear stresses through the bond between the soil and reinforcement and through its tensile strength of reinforcement.

With increasing embankment heights, the spreading forces, and as a result, the tensile forces on the reinforcement will be significantly increased and lead to higher deformations in the system. Both the membrane effect (arching effect) and the

D. Choudhury et al. (Eds.): GeoMEast 2018, SUCI, pp. 53–61, 2019.
https://doi.org/10.1007/978-3-030-01920-4_5

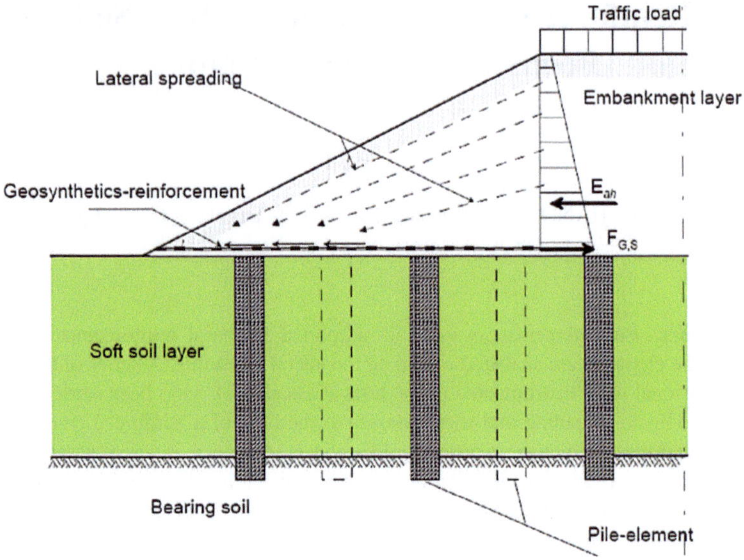

Fig. 1. Lateral spreading of a reinforced embankment

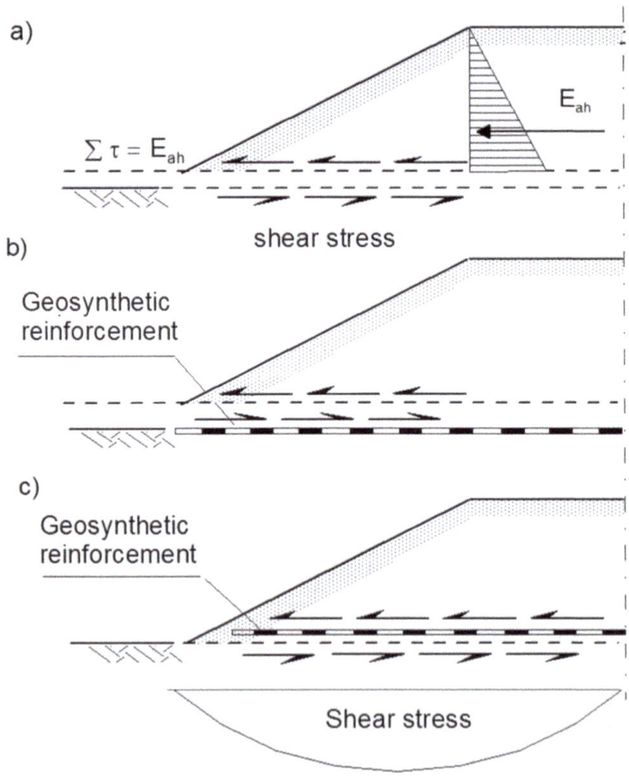

Fig. 2. Reinforcement in an embankment on soft underground a) Spreading forces and shear stress at base b) Reinforcement sustains the stresses; c) Effect of stress on underground

spreading effect influence the behavior of the bearing system (pile-like elements) and the tensile forces on the reinforcement. Therefore, a precise study to analyze and evaluate these effects for higher embankments has been carried out.

Moreover, the behavior of soil-reinforcement interaction must be accurately described in order to attain the real stress-strain relations in such zones.

The calculation of shear stresses and horizontal deformations at the embankment base, as well as the tensile forces in the geosynthetics reinforcement, has been determined through a series of large-scale model tests under variation of underground conditions. Similarly, the horizontal force on head of pile element due to spreading effect has also been measured and analyzed. The test-results have been verified using a finite element method. The goal of the validation processes is to calibrate the soil parameters obtained from laboratory tests for further FE-parameter study.

An extensive parameter study has been carried out by means of keeping one or more parameters constant and varying the others. The parameter study gives a clear overview of all the factors that influence the tensile forces in the reinforcement.

A modified analytical method has been introduced to estimate analytically the spreading force sustained by the reinforcement. The modified method can be used for qualitative determination of the spreading forces in the reinforcement in the case of high and very high embankments, different underground conditions, and different embankment slopes.

2 Large-Scale Model Tests

A section of an embankment was simulated in a model test at a scale of 1:3. Thereby, three different underground conditions were considered. These are:

a. Underground material the same as embankment (sand),
b. Soft underground represented by foam material, and
c. Soft underground with pile-like elements.

Each case was tested with and without a base reinforcement (geogrids).

The first model test was carried out on homogeneous sand without a surcharge load and it is aimed to investigate the effect of slope variations (1:3, 1:2 and 1:1.5) on the outward shear stresses at the embankment base. The shear stresses due to lateral spreading are derived from horizontal earth pressures measured at different sections in the slope zone taking equilibrium condition of an infinite vertical slice as follows:

$$\tau = \frac{\partial E_h}{\partial y} \tag{1}$$

Figure 3a shows the development of shear stresses in the slope zone at the embankment base under different slope variations. (B is the slope width, i.e. at embankment shoulder y/B = 1).

It can be seen from Fig. 3a that the steeper the embankment, the higher the shear stresses. The rest four model tests were conducted on loaded embankment with a slope of 1:1.5; with/without base reinforcement, and with/without pile-like elements (in a square grid).

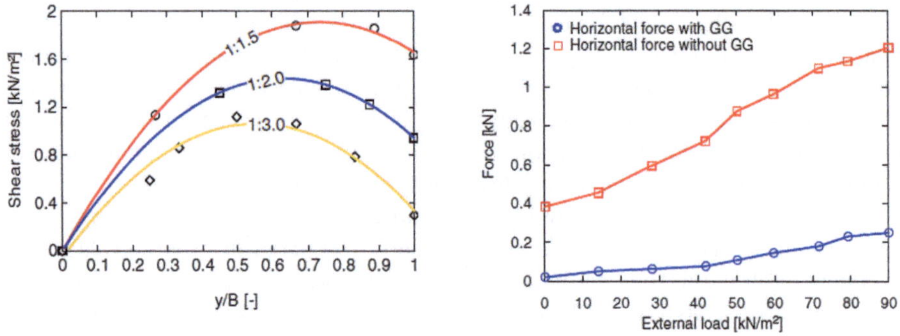

Fig. 3. a (Right). Shear stresses at embankment base under slope variations, b (left). Measured horizontal force in the pile-like element

The effect of the geogrid reinforcement has also been evaluated by measuring the horizontal force on the top of the pile-like elements with/without base reinforcement. It can be seen from Fig. 3b that the reinforcement reduces the horizontal force on pile head. For details of model test results and interpretations, see Fahmy (2008).

3 Verification of the Model Test Results

The model tests are verified using FEM. The soil parameters are first calibrated on the homogeneous model test (sand underground without reinforcement and pile-like elements). Comparison of the measured and calculated deformations at different sections and earth pressures in the slope zone shows good agreement (see Fahmy 2008) without major adjustment of the soil parameter. The rest of the models are then analyzed using the calibrated soil parameters. Plane strain FE-models were used to analyze the model tests of un-/reinforced embankments on un-piled soft underground, whereas three-dimensional FE- models had been employed in the case of a piled soft underground. Figure 4 shows selective results of the FE-computation compared to measured values. Figure 4a demonstrates that the calculated and measured strains in the geogrids agree very well in the case of underground without pile-like elements. Whereas, the calculated strains in the base reinforcement on top of a pile-like element shows a large difference (Fig. 4b). This may be attributed to the simulation of the geogrids as a membrane. The geogrid seems to behave differently as a membrane, especially when it is set on a point support system. This performance is reported by Zäske (2001), Bussert et al. (2004), Jenck et al. (2005), Heitz (2006), and Van Eekelen (2015). Based on the authors own results, such as shown in Fig. 4b, and back analysis of model test results from the

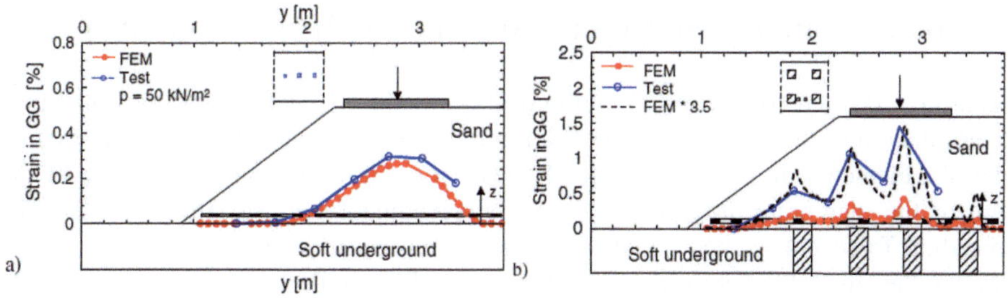

Fig. 4. Tensile strain in reinforcement a) un-piled and b) piled embankment.

literature (for Zäske 2001, Heitz 2006) a factor of about 3.5 is derived between calculated and measured results.

4 Parameter Study

Series of numerical parameter studies on the prototype using 3D-FEM are performed under different parameter variations such as embankment height, slope, underground stiffness, geogrids stiffness, and number of layers. The result of the numerical study also confirms that under steeper slope the shear stress at the slope base is greater, and consequently the resulting spreading forces are greater (Fig. 5). The effect of the slope is more evident by high embankments than the lower embankments. Both spreading and membrane forces in geogrid are also observed to be smaller in the case of stiffer underground than soft underground. This is attributed to the small shear deformations of the stiffer underground.

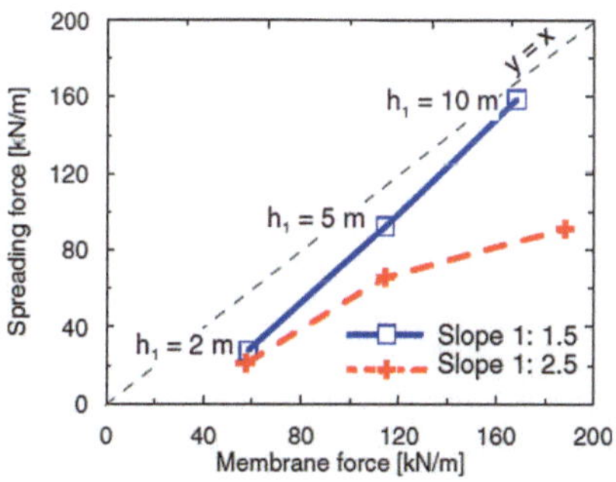

Fig. 5. Spreading and membrane forces under different embankment heights and slopes.

5 Comparison of Fem-Results with the Analytical Approach

EBGEO (2007) recommends two approaches (option 1 and 2) for the analytical determination of the total tensile force in reinforcement for embankments supported by pile-like elements. In option 1 the total tensile force in reinforcement is taken as the sum of the membrane force ($F_{G,M}$) and the spreading force ($F_{G,S}$) (see Eq. 2). The spreading force is assumed equal to the horizontal earth pressure force at a section through the crest of the embankment. This approach is similar to that recommended by BS 8006 (1995).

$$F_G = F_{G,M} + F_{G,S} \tag{2}$$

Option 2 is similar to the approach by Love and Milligan (2003). It is based on the concept that basal reinforcement can only have one tension in the transverse direction of embankment. The reinforcement in this case should be designed for whichever is the greater: the membrane force or spreading force (Eq. 3), but not their sum. This approach was also adopted by Klobe (2007).

$$F_G = max\begin{cases} F_{G,M} \\ F_{G,S} \end{cases} \tag{3}$$

Option 2 however requires that the piles should not sustain bending due to deflection of the pile heads. Figure 6 shows a comparison between the analytical and the numerical results in case of stiffer underground with flatter embankment slope 1:2.5.

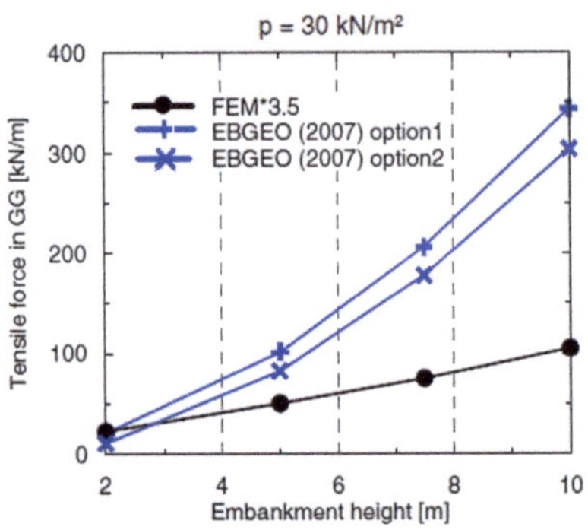

Fig. 6. Tensile forces in reinforcement determined analytically (EBGEO 2007) and numerically (FEM)

Figure 6 shows that both EBGEO (2007) options would lead to an overestimated tensile force in the reinforcement as compared to FEM-results, especially in the case of high embankments. This is mainly referred to the assumption that the spreading force is equal to the horizontal active earth pressure force at a section through the crest of the embankment disregarding the effect of underground stiffness.

6 Modified Analytical Method

A modification of the analytical method to calculate the spreading force is carried out based on the assumption that the section (through which the horizontal earth pressure force is determined) is not always fixed at the embankment crest, rather it moves towards the toe depending on the embankment height as shown in Fig. 7. The position and the height of the fictitious wall h_w depends mainly on a vertical angle θ from the slope crest. The critical angle θ can be determined by equating the so calculated spreading force with that obtained from FEM.

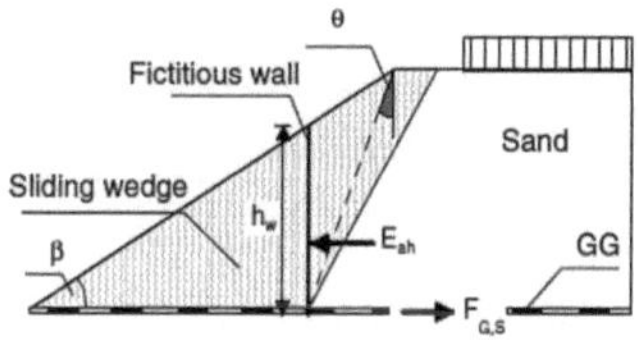

Fig. 7. Sliding soil wedge to determine the spreading force

The earth pressure force E_{ah} on the fictitious wall which is assumed equal to the spreading force $F_{G,S}$ can be calculated using the earth pressure theory.

By comparing the tensile forces due to spreading determined analytically and numerically, the critical angles for different embankment heights are identified for the case of a peat underground ($Es = 0.8MN/m^2$) and embankment slope of 1:1.5 (reference model). For embankment height up to 5 m, the critical angle is found to be $\theta = 0°$ and for $h_l = 10$ m; $\theta = 30°$.

For other underground condition and different embankment slope, adjustment factors are introduced to the tensile force due to spreading as shown in Eq. 4.

$$F_{G,S} = E_{ah}(h = h_w).f_{E_s}.f_\beta \tag{4}$$

Where:

E_{ah} $(h = h_w)$ is the earth pressure force for the reference model,
f_{Es} is dimensionless factor to account for the influence of the stiffness of the underground (f_{Es} = 1.0 in the case of peat underground), and
f_β is a factor to account for the influence of the slope of the embankment (f_β = 1.0 in the case of slope 1:1.5).

The factors f_{Es} and f_β can be formulated as functions of underground stiffness E_s and embankment slope β as follows:

$$f_{E_s} = k_1.E_s^{(k_2)} \text{ And } f_\beta = k_3.\left(\frac{1}{\tan\beta}\right)^{k_4} \tag{5}$$

Where κ_1, κ_2, κ_3 and κ_4 are variables that depend on embankment height h_1. Equation 4 can now be written in general form as:

$$F_{G,S} = E_{ah}.(0.92 - 0.0085.h_1).E_s^{(-0.4768-0.0165.h_1)}.(0.962 + 0.0504.h_1).\left(\frac{1}{\tan\beta}\right)^{(0.0334-0.099.h_1)} \tag{6}$$

Figure 8 is a comparison of the tensile forces in the reinforcement due to spreading according to the modified and EBGEO (2007) approaches for different slopes and underground conditions.

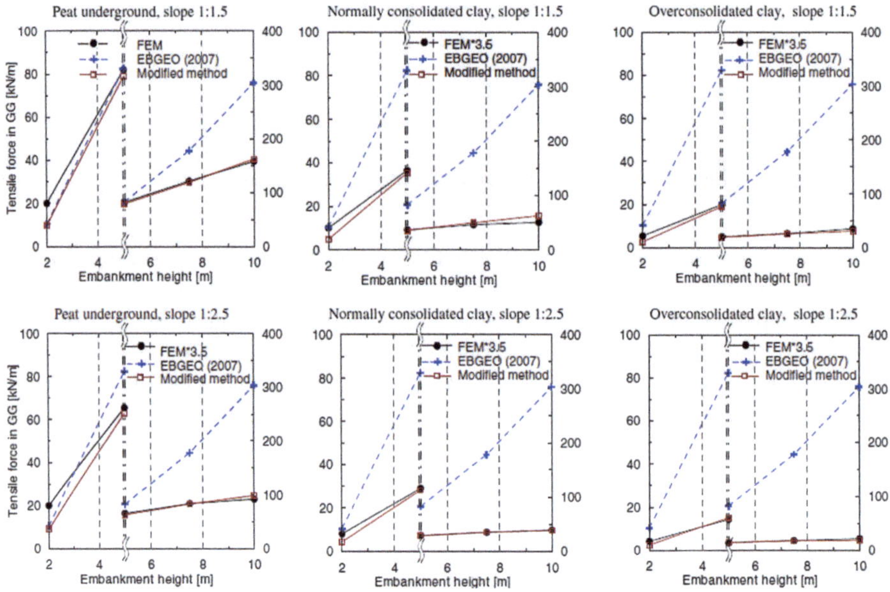

Fig. 8. Comparison of tensile forces in reinforcement due to spreading according to the EBGEO (2007) and the modified approaches

7 Conclusion

Conventional analytical methods to determine the spreading forces are overestimated, especially for high embankments, as these methods ignore the effect of the embankment slope and the underground stiffness. The modified method based on model tests and intensive FE-parameter study takes into account all these effects and leads to a reasonable tensile force in the geogrid basal reinforcement.

References

Bussert, F., Meyer, N., De Lange, A.P., Van der Stoel, A.E.C.: Bemessung und Üb-erwachung einer geokunststoffbewehrten Tragschicht auf HSP-Pfählen. Bautechnik Heft **81**(12), 959–967 (2004)

EBGEO Empfehlung für den Entwurf und die Berechnung von Erdkörpern mit Bewehrung aus Geokunststoffen, Ausgabe 2007 (2007)

Fahmy, G.S.: Lateral spreading in basal reinforced embankments supported by pile-like elements. Shriftenreihe Geotechnik, Universität Kassel, Heft 20 (2008)

Heitz, C.: Bodengewölbe unter ruhender und nichtruhender Belastung bei Berücksich-tigung von Bewehrungseinlagen aus Geogittern. Schriftenreihe Geotechnik, Universität Kassel, Heft 19 (2006)

Jenk, O., Dias, D., Kastner, R.: Soft soil improvement by vertical rigid piles. In: 2nd International Workshop of Young Doctors in Geomechanics, ENPC, Champs-sur-Mame (2005)

Kempfert, H.-G., Stadel, M., Zaeske, D. eds.: Berechnung von geokunststoffbewehrten Tragschichten über Pfahlelementen. Bautechnik Heft **75**(12), 818–825 (1997)

Love, J., Milligan, G.: Design methods for basally reinforced pile-supported embankments over soft ground. Ground Eng. **36**(3), 39–43 (2003)

Van Eekelen, S.J.M.: Basal Reinforced Piled Embankment, experiments, field studies and the development and validation of a new analytical design model, TU-Delft (2015)

Zäske, D.: Zur Wirkungsweise von unbewehrten und bewehrten mineralischen Tragschichten über pfahlartigen Gründungselementen. Schriftenreihe Geotechnik, Universität Kassel, Heft 10 (2001)

Experimental Study on Lateral Resistance of Steel Pipe Pile with Wings Using the New Composite Geo-material with Magnesium Acrylate

Atsushi Shimamura[1(✉)], Yuji Miyamoto[2], and Toshio Kobayashi[2,3]

[1] Engineering Development Department, Chemical Grouting Co., Ltd., Tokyo, Japan
a-shimamura@chemicalgrout.co.jp
[2] Department of Architectural Engineering Division of Architecture, Osaka University, Osaka, Japan
[3] Sansei Inc., Tokyo, Japan

Abstract. In recent earthquakes in Japan, there are many cases that damages on soil-foundation systems negatively affect the functions of the superstructures. It has become important from this point to improve the aseismic capacity of soil-foundation systems.

Soil improvement is one of the measures against ground damages, and "soil-cement mixture" is the material often used, but it sometimes exhibits brittleness, and results in decreasing the aseismic capacity.

Steel pipe piles with wings have high deformation performances, high vertical bearing capacities, and high pull-out resistances. However, these piles loosen soil around the pile shafts, and the loosened soil may affect the lateral subgrade reaction of the piles.

Authors developed an artificial soil having toughness against large deformation.

This paper discusses laboratory studies of the shearing mechanism and the properties of newly developed composite geo-material consisting of magnesium acrylate. Then, report the results of static cyclic loading tests in the actual ground that was to investigate lateral resistance of the steel pipe piles with wings strengthened by the composite geo-material.

1 Introduction

In recent earthquakes in Japan, there are many cases that damages on soil-foundation systems, such as damages on piles and footing beams caused by ground liquefaction or settlement around foundations, negatively affect the functions of the superstructures [1]. There were cases that those damages, such as inclination in buildings by damaged piles and differential settlement of detached houses by ground liquefaction in the 2011 off the Pacific coast of Tohoku earthquake. Therefore, it has become important to improve the aseismic capacity of the soil-foundation systems.

Soil improvement is one of the measures against those ground damages, and "soil-cement mixture" is the material often used [2, 3], but it sometimes exhibits brittleness with the increase in the strength, and results in decreasing the aseismic capacity.

D. Choudhury et al. (Eds.): GeoMEast 2018, SUCI, pp. 62–74, 2019.
https://doi.org/10.1007/978-3-030-01920-4_6

Steel pipe piles with high deformation performance are excellent in aseismic capacity. Among them, steel pipe piles with wings attached to the tip of the piles have especially high vertical bearing capacities and high pull-out resistances because the piles are driven by the wings into the ground by rotation. However, soil around the pile shaft is loosened because the wings are larger than the pile diameter, and the loosened soil may affect the lateral subgrade reaction of the pile.

To address this problem, it is important to develop an artificial soil having toughness against large deformation with clear physical properties, then to understand the behavior of the soil-foundation system during a large earthquake, and to enhance the aseismic capacity of the superstructure and the soil-foundation system as a whole. In this way, the aseismic capacities of the structures are improved and the functions of the structures are maintained even after events of large earthquakes.

Focusing on the ductility and the elasticity of magnesium acrylate, authors have developed an artificial geo-material compound, which is a mixture of soil and magnesium acrylate and stable in a large strain region. The composite geo-material is used as a bearing stratum for buildings, backfill soil around foundations and improves soil around piles [4, 5] etc.

This paper discusses laboratory studies of the shearing mechanism and the properties of the newly developed composite geo-material consisting of magnesium acrylate. Authors also performed the unconfined compression test and the cyclic simple shear test in laboratory, then static cyclic loading test in actual ground for investigating lateral resistance of the steel pipe piles with wings strengthened by the composite geo-material around the pile head.

2 Summary of the New Composite Geo-Material

This paper discusses laboratory studies on the shearing mechanism and the properties of the new composite geo-material including unconfined compression test and the cyclic simple shear test.

Table 1 shows the physical properties of the material used in this study. It is formulated with sodium thiosulfate (reductant), ammonium peroxodisulfate (oxidant), and magnesium acrylate solution as a base compound. The gel time of the geo-material is from 1 to 2 h. The peak strength develops after 3 days, and then mechanical properties become stable.

Figure 1 illustrates the curing mechanism of the geo-material. The acrylic acid salts polymerize into chain structures then develops further the insoluble network structures of the three dimensions [6]. Inside the new geo-material, not only Magnesium acrylates polymerize and fill voids within sand particles, but, as an electrolyte, also chemically combine with sand particles, making the geo-material become firm solid. The volume ratio of Chemical to Soil is 1 to 1.44.

Table 1. Material properties

Composite material	Description
Chemical	(a) Magnesium acrylate solution (29 wt%) Density = 1.17 g/cm^3
	(b) Sodium thiosulfate Density = 1.67 g/cm^3
	(c) Ammonium peroxodisulfate Density = 1.98 g/cm^3
Soil	Silica sand (Tohoku silica sand No. 6) Density = 2.58 g/cm^3 Particle size range = 0.075–0.425 mm

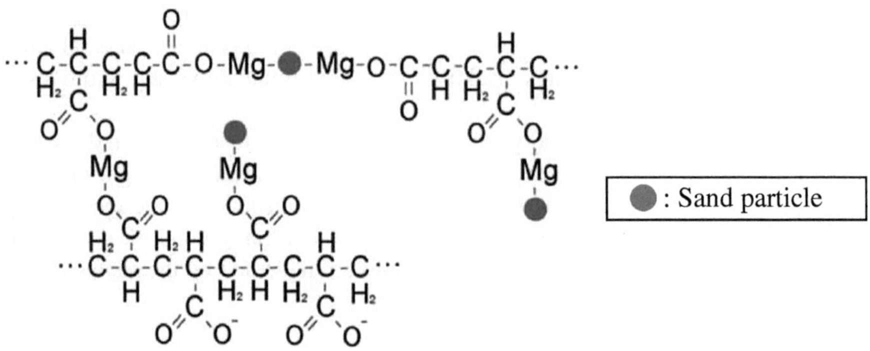

Fig. 1. Curing mechanism of the new composite geo-material

3 Laboratory Tests

3.1 Unconfined Compression Test

According to the Standard of Japanese Geotechnical Society (JGS) [7], the unconfined compressive strength, q_u, is defined as the compressive stress at which an unconfined cylindrical specimen of soil will fail in a simple compression test. The test started by applying a constant axial strain of 1% per minute. The load and deformation values were recorded as needed for obtaining a reasonably complete load-deformation curve. Samples used for the test were within 3 days old after preparation. Specimens were 5 cm in diameter and 10 cm in length. We defined the strain at the maximum compressive stress (unconfined compressive strength) q_u, as the failure axial strain, ε_f, on these tests.

3.2 Cyclic Simple Shear Test

As shown in Fig. 2, cyclic simple shear testing apparatus used in this study is of advanced Kjellman-type [8]. The testing apparatus was adopted in order to accurately evaluate the shear modulus, G, and damping ratio, h, over a wide range of shear strain, γ, from 0.001% (10^{-5}) to that at the peak of the order of 10% (10^{-1}). A cylindrical specimen, which was laterally confined by Teflon coated low friction rings, was subject to a cyclic simple loading at a small level. The relation between shear stress, τ, and shear strain, ε, was determined. The loading method was strain-controlled. The sine wave loading of 11 cycles was applied continuously for each step in an undrained condition, and the vertical stress, σ_v, was 30–400 kN per square metre. We adopted loading frequency at 0.1 Hz. Samples were used for the test within 3 days after preparation. Specimens were 10 cm in diameter and 4 cm in length. The samples were fixed to upper loading plate and shaking table with epoxy resin adhesive. We defined the shear modulus at very small strain level ($\gamma \approx 0.001\%$) as the initial shear modulus, G_0, on these tests.

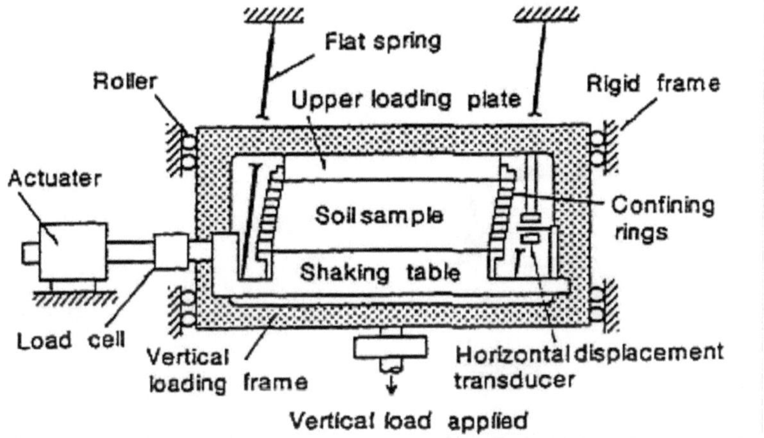

Fig. 2. Schematic of cyclic simple shear testing apparatus

4 Results of Laboratory Tests

4.1 Unconfined Compression Test

Figure 3 illustrates the relations of axial strain, ε, and compressive stress, σ, of the new composite geo-material and usual soil-cement mixture. It indicates that the new composite geo-material has higher strain than that of usual soil-cement mixture with the strength of the same degree. Soil-cement mixture also shows high brittleness from the relation between axial strain and compressive stress. The failure axial strain of the soil-cement mixture sample was about 1%, and when 1% was exceeded the specimen was brittle fractured as shown in Photo 1b. However, the failure axial strains have increased about 13% with the addition of the new composite geo-material (sample No. 1–3) which is the mixture of Chemical and soil. The deterioration of strength of composite geo-material in a high strain region after peak strength was small, and the shape of the specimen was maintained even in the region of over 15% as shown in Photo 1a. Three samples of the same composition (sample No. 1–3) indicated the same characteristics.

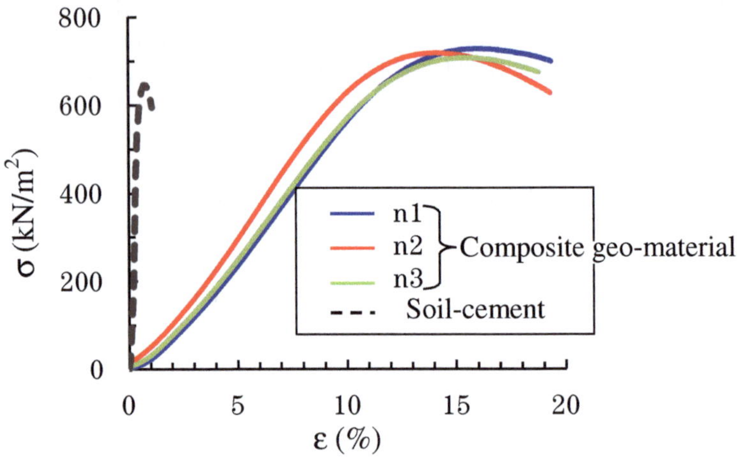

Fig. 3. Stress-strain curves

a) **Composite geo-material (ε>15%)** b) **Soil-cement mixture (ε<2%)**

Photo 1. Failure conditions of specimens after unconfined compression test

4.2 Cyclic Simple Shear Test

Figure 4 shows the relations between shear modulus, G/G_0, and shear strain, γ, of the new geo-material and usual soil-cement mixture. In each sample, shear moduli decreased nonlinearly with increasing shear strain for all vertical stresses. The decreasing in shear modulus of the composite geo-material with increasing shear strain was smaller than those of the soil-cement mixture in the very large strain range. That is because the maximum shear stress of the composite geo-material increased also in the very large strain range (Fig. 5). Figure 6 shows the relations between initial shear modulus, G_0, and vertical stress, σ_v, of the new composite geo-material and usual soil-cement mixture with the compressive strength of the same degree. In each sample, shear moduli decreased nonlinearly with increasing shear strain for all vertical stresses.

The initial shear modulus increased with increasing vertical stress. The initial shear modulus of the composite geo-material was smaller than that of the soil-cement mixture for the same vertical stress.

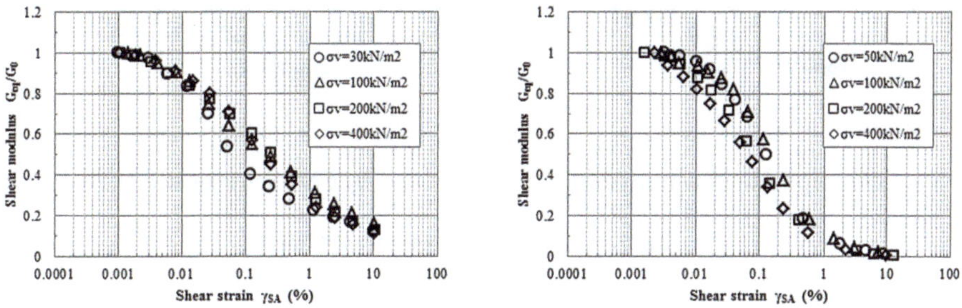

a) Composite geo-material **b) Soil-cement mixture**

Fig. 4. Shear modulus vs. shear strain

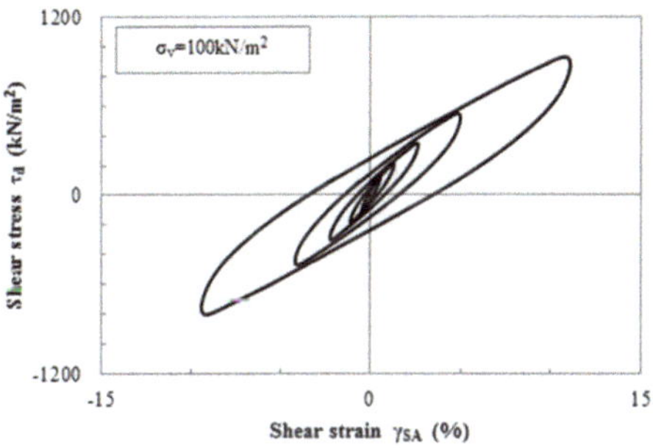

Fig. 5. Shear stress vs. shear strain (Composite geo-material)

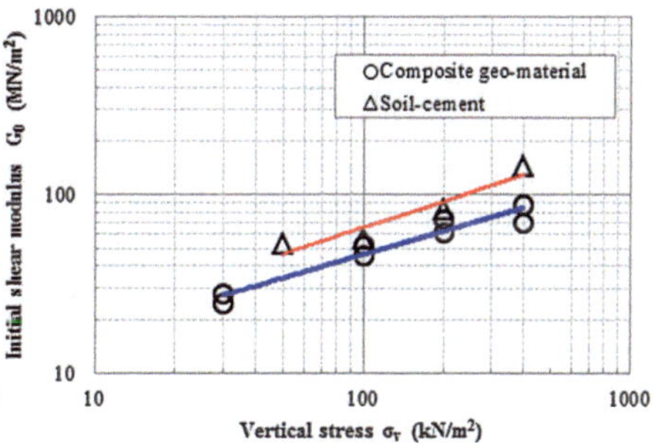

Fig. 6. Initial shear modulus vs. vertical stress

Figure 7 shows the relation between damping ratio, h, and shear strain of the new composite geo-material with the result of the usual soil-cement mixture at the compressive strength of the same degree. As for each sample, damping ratios increase nonlinearly with increasing shear strain for all vertical stresses. As for the composite geo-material samples, damping ratios reached in the very large strain level of about 15–20% in, whereas damping ratios of soil-cement mixture showed large values in the large strain range ($\gamma > 3\%$). We consider that the specimens of the soil-cement mixture were fractured in the super-large strain range, so that the friction increased in the failure planes.

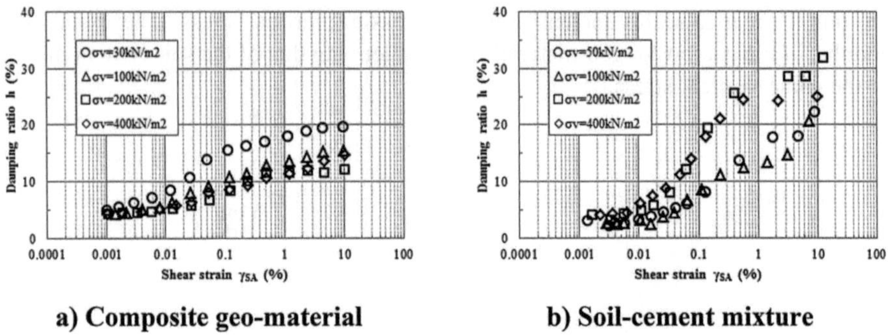

Fig. 7. Damping ratio vs. shear strain

5 Static Cyclic Loading Tests in Actual Ground

Lateral resistance of the steel pipe piles with wings strengthened by the new composite geo-material around the pile heads were investigated by static cyclic loading test in the actual ground.

5.1 Experiment Site

This experiment was conducted in Omitama City, Ibaraki Prefecture. Figure 8 shows the N value distribution, the soil log, and the main soil properties for the ground of the test site. The soil types are; from the ground surface to 0.55 m below ground level (bgl) is andosol, from 0.55 m bgl to 2.2 m bgl is weathered volcanic ash, from 2.2 m bgl to 3.2 m bgl is volcanic ash, and beyond 3.2 m bgl is sand.

5.2 Experiment

For the experiment, a JIS G 3444 general structural carbon steel (STK) pipe with the outer diameter φ 267.4 mm and the material thickness t 9.3 mm (steel grade 490) was used. The yield bending moment M_y of the pile is 148.1 kN·m. The pile with wings was fabricated by welding two flat plates to the tip of the steel pipe such that the inclination angle of each wing became 10° (Photo 3). A triangular excavation blade having a size to fit in the inner diameter of the steel pipe was attached by welding to the tip of a pile without wings (hereinafter, ST) (Photo 2). The tips of all test piles 5 m in length were blocked. To measure the bending strain of the pile on the inner surface of the pile shank during the experiment, strain gauges were attached at 50 points per surface in the loading direction at 100 mm intervals.

Figure 9 shows the experimental cases. A total of three test piles were installed. One ST was installed and used as a reference test pile. Two piles with wings (W23) were introduced with a wing diameter ratio Dw/Dp ≈ 2.3. One of them was strengthened by composite geo-material around the pile head (W23I). Construction range of the composite geo-material was 1.0 m (1.0 × 1.0 × 1.0 m) around the pile head. Curing period was more than 3 days.

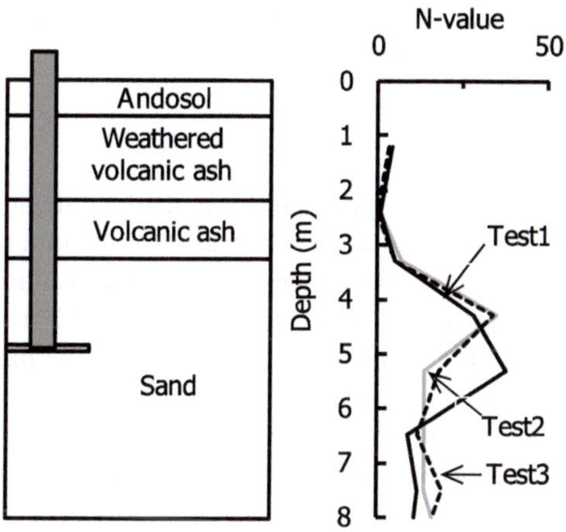

ρ (t/m³)	c (kN/m²)	φ (°)	Vs (km/s)	ν
1.28	29.3	16.9	0.13	0.29
1.34	35.1	10.6	0.10※	0.44※
1.64	89.6	9.3	0.16	0.47
1.80	1.80	1.80	0.15※	0.47※

Fig. 8. Soil properties of the test site

Photo 2. Test pile without wings (ST)

Photo 3. Test pile with wings (W23)

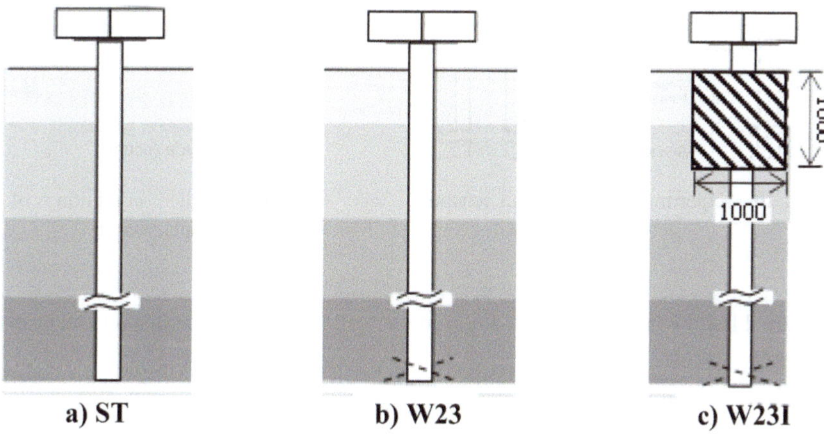

Fig. 9. Experimental cases

The test piles and reaction piles were installed by rotation, with the pile heads protruded at GL+200 mm above ground level (GL) when completed.

5.3 Loading Method

Figure 10 shows an overview of the experimental apparatus. The loading device consisted of a test pile, reaction piles, and a hydraulic jack. A base plate with 25 mm in the thickness was welded to the head of the test pile and reaction piles in order to attach a load beam and other components. Four steel pipes of φ 406.4 mm × t 12.7 mm (material STK490) were used as the reaction piles. Against the reaction force, the pile heads of the four reaction piles were combined in a square form with H-shaped steel. The hydraulic jack was attached between the load beam and the reaction block, which were joined by high strength bolts to the test pile, and the experiment was conducted under conditions allowing free rotation of the head of the test pile. The loading height was at GL+400 mm. In addition to pile strain, the lateral displacement and out-of-plane displacement of the test pile and reactive body were measured at the loading height position with a displacement meter and laser displacement meter, at the center of the pile head. Then an inclinometer was installed on the center of the head of the test pile and the reactive body to measure the tilt angle. The lateral displacement of the test pile was measured at two points: the load height and at GL+100 mm. The lateral displacement at the position of GL+100 mm is referred to as "reference displacement".

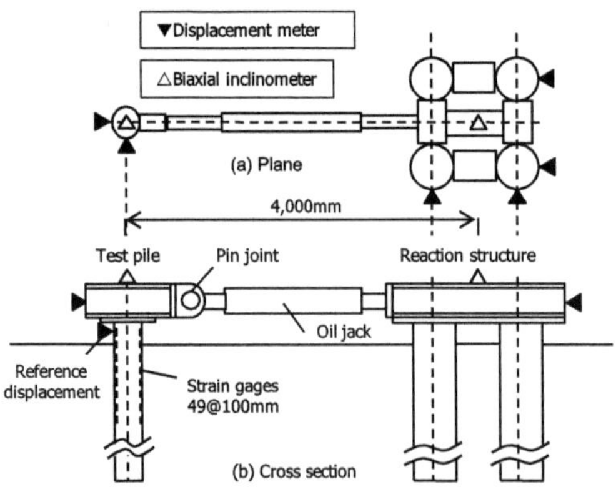

Fig. 10. Experimental device and sensors

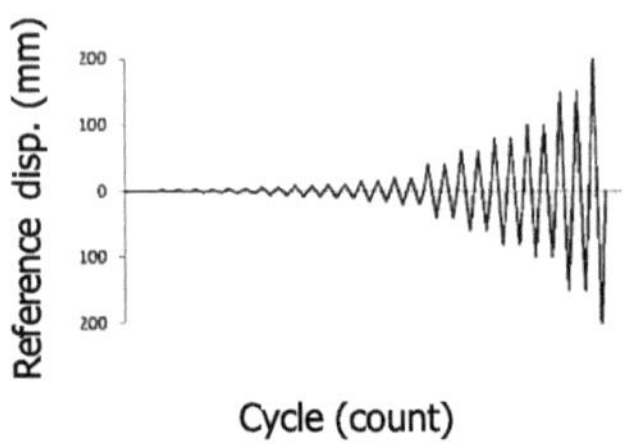

Fig. 11. Cyclic condition of Reference displacement

Figure 11 shows the load cycle diagram controlled by reference displacement in the experiment.

6 Results of Static Cyclic Loading Tests in Actual Ground

6.1 Pile Head Load vs. Reference Displacement

Figure 12 shows the relation between the pile head load and the reference displacement. The mark (□) was the point that axial strain of the pile reached yield strain (1500 μ) in the figure. The initial loading curve showed a decrease in rigidity from the start of loading. In addition, as the reference displacement increased, the loop properties became an inverted S-shape, indicating slip. This is presumed to be caused by plastic deformation of the ground from the start of the loading. The pile deformation increased and the plastic deformation of the ground increased.

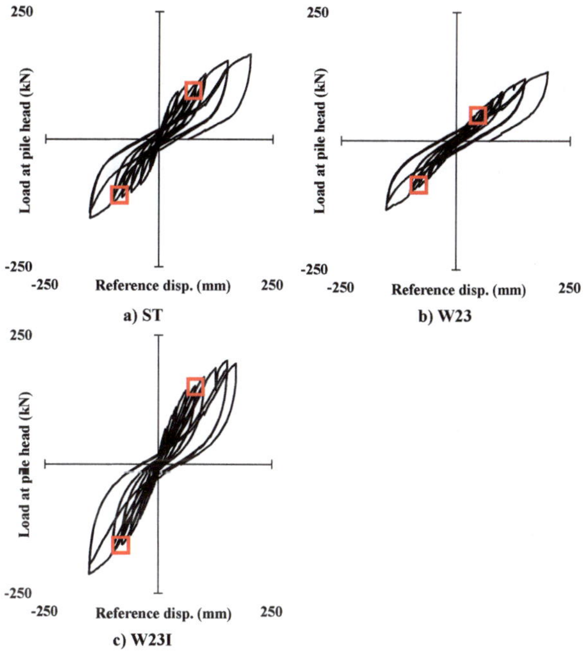

Fig. 12. Pile head load – reference displacement

6.2 Ground Deformation

Figure 13 shows the ground deformation. The figure shows the cracks on the ground surface at the peak load in the first cycle with deference in displacement amplitudes of 150 mm. Many cracks were found for ST in front and behind the piles. There were fewer cracks for W23 compared with ST. For W23I, cracks were found on the boundary face between the original ground and the composite geo-material. Cracks were also found in the composite geo-material in the direction perpendicular to the loading. In both experiments, detachment between piles and soil foundation were observed.

6.3 Characteristics of Bending Moment Distribution

Figure 14 shows the relation between the pile head load and the ground surface reference displacement. Figure 15 shows the bending moment distribution in the depth direction for the reference displacements of 20 mm and 60 mm. Here, the bending moment was calculated from the strain gauges on the piles. At the same ground surface position reference displacement, the pile head load of W23I is larger than those of W23 and ST (Fig. 14). Also, the maximum bending moment of W23I is larger than W23 when being displaced identically regardless of pile displacement, and its occurrence depth becomes shallow (Fig. 15). Though there is a difference in the pile top load of W23I in comparison with ST, the bending moment distributions are of little difference. This is regarded as because 1 m shallow soil improvement around the pile head generates the large foundation resistance.

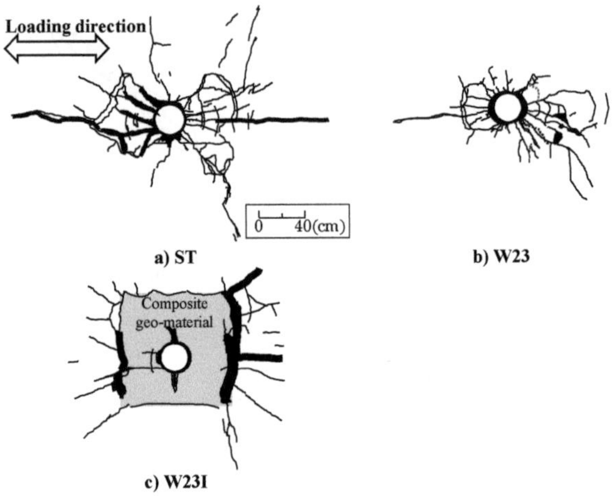

Fig. 13. Cracks of the ground surface around test pile

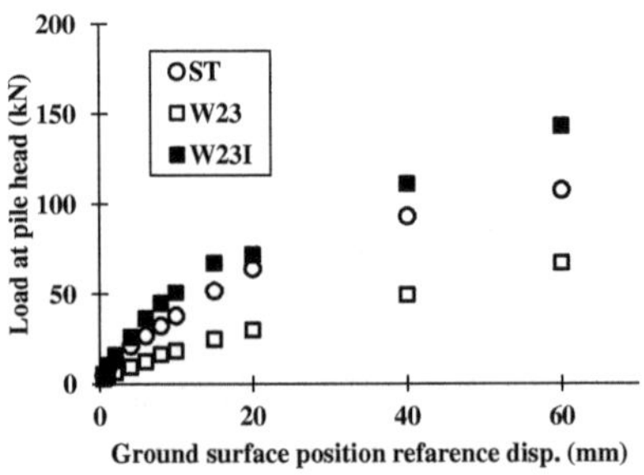

Fig. 14. Pile head load – the ground surface reference displacement

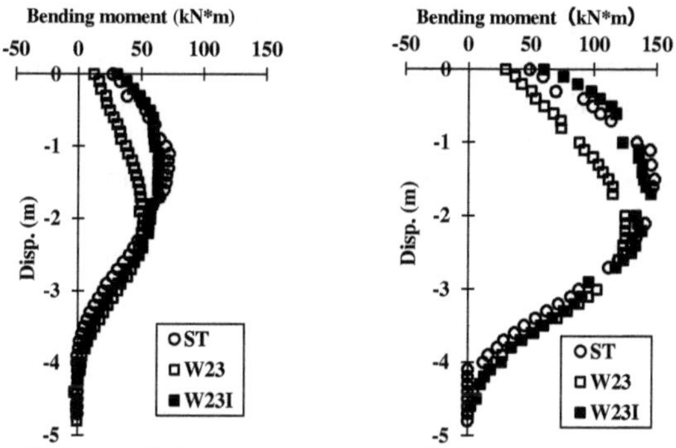

a) Reference displacement at 20mm cycle

a) Reference displacement at 60mm cycle

Fig. 15. Bending moment distribution

7 Conclusions

In this research, we have studied the properties of newly developed composite geo-material with magnesium acrylate on laboratory studies, and the lateral resistance of steel pipe piles with wings strengthened by the composite geo-material around the pile heads on the static cyclic loading test. The findings of the present study are as follows.

(a) The composite geo-material with magnesium acrylate has a high toughness compared with a usual soil-cement mixture base on laboratory tests.
(b) In the static cyclic loading test in the actual ground, authors confirmed that the pile head load and the bending moment of the steel pipe pile with wings strengthened by new composite geo-material around the pile head were increased at the same ground surface position reference displacement, and that the reinforcing effect of soil-pile foundation system including composite geo-material.

Acknowledgments. This work was supported by JSPS KAKENHI Grant Number JP17H03342 from Japan Society for the Promotion of Science (JSPS). The authors express their sincere thanks to the above organization.

References

1. AIJ: Preliminary Reconnaissance Report of the 2011 Tohoku-Chiho Taiheiyo-Oki Earthquake. AIJ (2011). (in Japanese)
2. AIJ: Recommendations for Design of Building Foundations. AIJ (2001). (in Japanese)
3. The Building Center of Japan: Recommendations for Quality Control and Design of Ground Improvement for Buildings. The Building Center of Japan (2002). (in Japanese)
4. Shimamura, A., et.al.: Development of compound geo-material with magnesium acrylate and experimental study on the earthquake response reduction. In: Summaries of Technical Papers of Annual Meeting. AIJ (2016). (in Japanese)

5. Kobayashi, T., et.al.: A study on lateral resistance of screw pile with wing by lateral load test. In: Summaries of Technical Papers of Annual Meeting. AIJ (2017). (in Japanese)
6. Miyazawa, K., Kondo, K.: Acrylic acid salts type soil stabilizer. Extra edition No.10 Soil stabilizing polymer of Polymer applications, pp. 55–63 (1972). (in Japanese)
7. JGS: Japanese Geotechnical Society Standrds Laboratory Testing Standards of Geomaterials, vol. 1. JGS (2015)
8. Hara, A. and Kiyota, Y.: Dynamic shear tests of soils for seismic analysis. In: Proceedings of 9th ICSMFE (Tokyo), vol. 2, pp. 247–250 (1977)

Vibration Isolation of Foundation Using Hdpe and Natural Geocells - A Review

Sreevalsa Kolathayar(✉)

Department of Civil Engineering, Amrita Vishwa Vidyapeetham, Coimbatore, India
sreevalsakolathayar@gmail.com

Abstract. Geocells are three-dimensional, polymeric, honeycomb like structures which are interconnected by joints and are used as confinement of soil, gravel, concrete, etc. Geocell, is one of the types of geosynthetic, that provide all round confinement due to its three dimensional structure. The soil-geocell layers acts as a stiff mat and distribute the vertical loads over a wider area of the soil. In this paper, review on the application and impact of geotextile and geocells as reinforcement in the foundation bed under static and dynamic conditions are presented. The previous studies on performance of both natural and polymer based geosynthetics and geocells are presented in the paper. Based on the detailed literature review this paper lays out identified research gaps and presents future directions for research to explore the potential of geotextiles and geocells made of natural materials, for vibration isolation.

1 Introduction

Geosynthetics, one of the main techniques of soil reinforcement have been widely used as construction materials in civil engineering projects. Today, the geosynthetic products like geocell, geogrid, geotextile, geomembrane etc., are available in the market. Geosynthetics are used in locations where shear stresses are generated. The shear resistance between soil and Geosynthetics limits the lateral deformation of the soil. These are used for increasing bearing capacity and in reducing settlement of soil. Each product is designed to solve a specific type of civil engineering problems. The term geocell also have two parts. The first part, "Geo" means soil or earth and the second part "cell" represents the cellular type of shape for infill material such as soil, gravel, concrete, etc. It forms a stiffer mat like structure and distributes the load induced to wider areas.

The geocells are considered a cost effective and durable. Now-a-days, geocell applications are growing at a rapid rate due to its proven advantages over traditional techniques of soil reinforcement. Using a stiff cell wall to confine the infill would create a situation where the failure of the confined infill will occur only when the solid particles crush or the cell walls undergo large deformation or rupture. The geocell-reinforced base can provide bending, tensile, and shear resistances and intercept failure planes from the subgrade.

D. Choudhury et al. (Eds.): GeoMEast 2018, SUCI, pp. 75–86, 2019.
https://doi.org/10.1007/978-3-030-01920-4_7

2 Studies on HDPE and NPA Geocells Under Static Condition

Rajagopal, et al. (1999) conducted a series of triaxial tests on geocells made up of woven and non-woven polypropylene geotextiles. They found that the infill soil imparts cohesive strength even though it is cohesionless soil. This cohesive strength depends upon the tensile modulus of geosynthetic material used to form geocells. The stiffness of composite increase by providing geocell reinforcement. Minimum three interconnected geocells are required to understand the performance of interconnected geocells layer bed. With geocell, soil can withstand high tensile stresses. The frictional strength of both reinforced and unreinforced soil samples will remain same and the failure envelopes of these samples will be parallel to each other.

Chen et al. (2013) performed triaxial compression tests on high density polyethylene (HDPE) geocells of various shapes (circular, rectangular and hexagonal), sizes and number of cells to evaluate compression strength of samples and stress- strain behavior of soil samples. It was found that the shear strength of granular soil can be improved by the confinement effect of geocells. Due to the number of corners hexagonal cells showed the higher friction angle than circular and rectangular cells.

Pokharel et al. (2010) experimentally investigated the stiffness and bearing capacity of single geocell-reinforced bases, including shape, type, embedment, height of geocells, and quality of infill materials. Four types of geocells (HDPE geocell and three Novel Polymeric Alloy (NPA) geocells with different tensile strengths) were used with Kansas River sand and quarry waste as infill materials. The elliptical shaped geocells failed suddenly and the infilled soil in it escaped from the geocell from the bottom of it. Due to this the confinement effect of geocell was lost. But, this loss was not seen in circular shaped geocells as they fail gradually. The stiffness and bearing capacity of reinforced base depends on elastic modulus of geocell sheet. The geocell without infill material had shown poor performance due to the breakage of weld under the application of load. The average stress was reduced and the load capacity was increased when there was an increase in the lateral expansion of cross section of the geocell. In the confined geocell, the deformation in the loaded cell was minimized and the rupture of the weld was immobilized due to the presence of the surrounding sand. The cohesion existing in the base material minimizes the benefit of the geocell for lateral confinement under static loading.

Kargar et al. (2017) described the influence of geocell geometrical properties on the improvement in bearing capacity and settlement of footing by loading strip footing model to the ultimate failure level. Compared to the unreinforced sand, the failure zones on the surface of reinforced soil were extended to a considerably larger distance beyond the edge of strip footing. The geocell mattress acted as a secondary foundation and it redistribute the footing stresses over a wider area in the bottom of the geocell cushion with a dispersion angle. The overall frictional resistance on the geocell wall, the moment of inertia and the bending rigidity of the geocell mattress increased, with increase in the height of the geocell wall. The lateral movements of soil particles inside the geocell will be restrained with the confinement of geocells. The membrane effect of the geocell reinforcement develops a tensile force in the reinforced bed and reduces the settlement. The points below the middle axis of geocell walls, under the footing center, undergo large strains up to the ultimate strain for geocells with h/B > 0.5 (h is the height of the geocell and B is the width of the footing). Consequently, a sudden shear failure takes place leading to a large heave in the soil surface beside the footing width and the infill soil of the geocell moves out of the pockets. As the diameter of geocell pocket increases, the number of cells under the loading area decreases leading to less confinement effect. Footing loads will be distributed over a larger area due to the rigidity of the geocell layer. This can be achieved by increasing geocell width. The extension of geocell width to two times of footing width in each side of the footing, provides anchorage from both sides of the loaded area due to the frictional and passive resistance developed at the soil-geocell interfaces.

3 Studies on Polymer Geotextile Under Static Condition

Tafreshi et al. (2016) manufactured a geocell and geotextile layers from the same geosynthetic material. The tests were conducted on these geocells to study their performance when circular footing is placed on it. The multiple geocell layers were laid under the footing. They observed that the first layer of geocell makes the stress field to pass into the deeper layers of the soil and the next two layers provide the tensile capacity to oppose the outward shear even under low settlements. Due to this reason, more than three layers of geocell did not show any effect. The geotextile layers act sequentially, whereas the geocell layers act at a time to resist the load. The same or greater improvement in vertical stiffness can be achieved by geocell layers than by planar sheets.

Dash et al. (2003) performed model footing tests on a circular footing supported on geocell made of polypropylene and observed that the encapsulated sand which was used as infill material owed to the confinement effect from the geocell walls and behaved as a relatively rigid member. This rigid member supports the footing load by distributing the frictional resistance over its outer periphery throughout the height of the geocell wall. By this it reduces the pressure on the underlying soft soil and increases sand bed performance. Surface heaving can be reduced by increasing the width of the geocell layer. This leads to the development of rupture planes in the soil bed by inducing a better composite behavior. The basal geogrid layer besides geocell further improves the performance of the footing both in terms of stiffness and load carrying capacity of the soil bed. The geogrid layer resists the downward deflection of the geocell mattress under footing load by its strength and membrane action. At higher settlements, the geocell layer deflects more and induces higher deformation in the geogrid layer at its base. By this action the geogrid layer gets strained more and hence mobilizes higher strength that gives rise to higher performance improvement. On the other side, an increase in the height of the geocell mattress increases the rigidity. This gives more uniform settlement at the base of the geocell mattress and reduces the contact pressure on the underlying soil. Load bearing capacity of planar reinforced soil reduces, when the soil between two successive layers gets squeezed out. In the case of geocell reinforcement the cells infilled with soil provide all-round confinement. This makes the geocell mattress behave as a composite body, even at larger footing settlement.

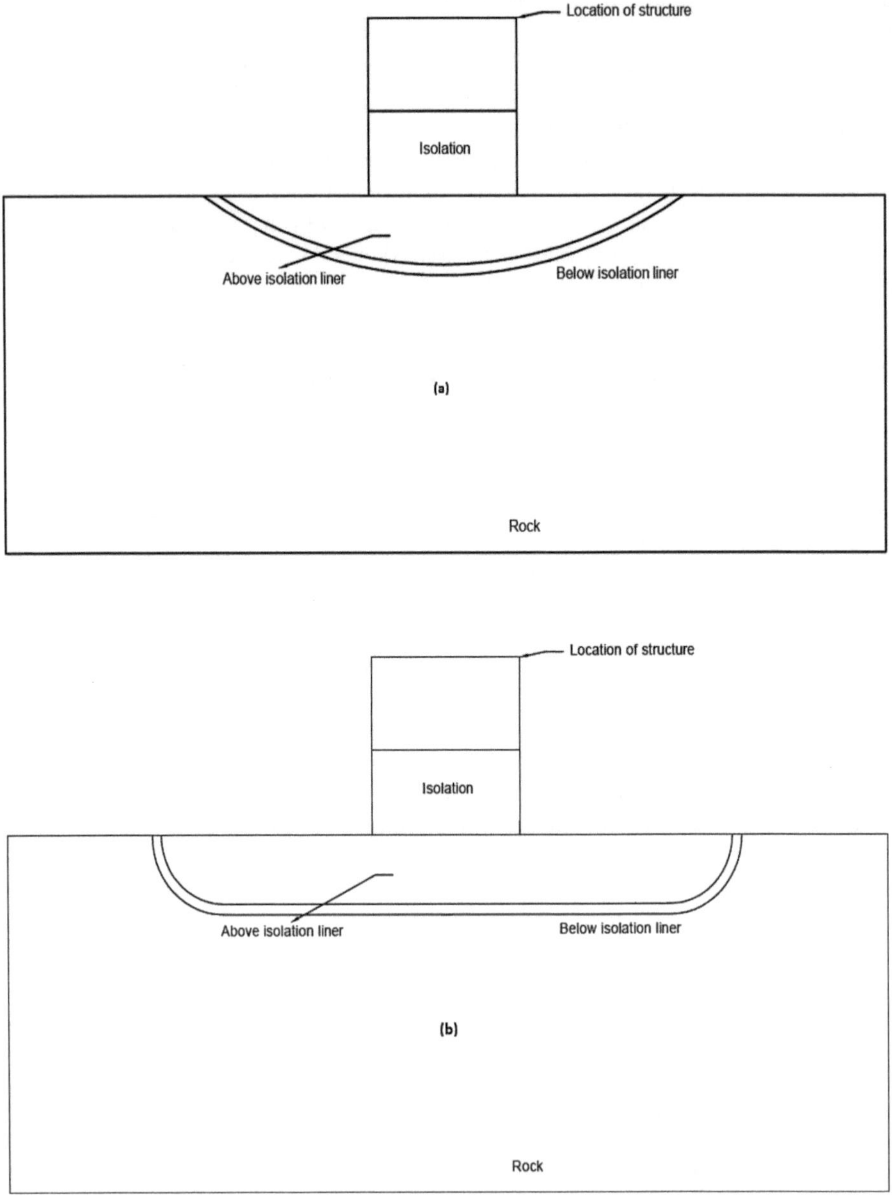

Fig. 1. In-ground isolation systems: (a) cylindrical liner geometry, (b) tub liner geometry, (c) trapezoidal liner geometry, (d) compound trapezoidal liner geometry, **Source**: Georgarakos, P., Yegian, M. K., and Gazetas, G. (2005), ***In-Ground Isolation Using Geosynthetic Liners***, 9th World Seminar on Seismic IsolationEnergy Dissipation and Active Vibration Control of StructuresKobe, Japan, June

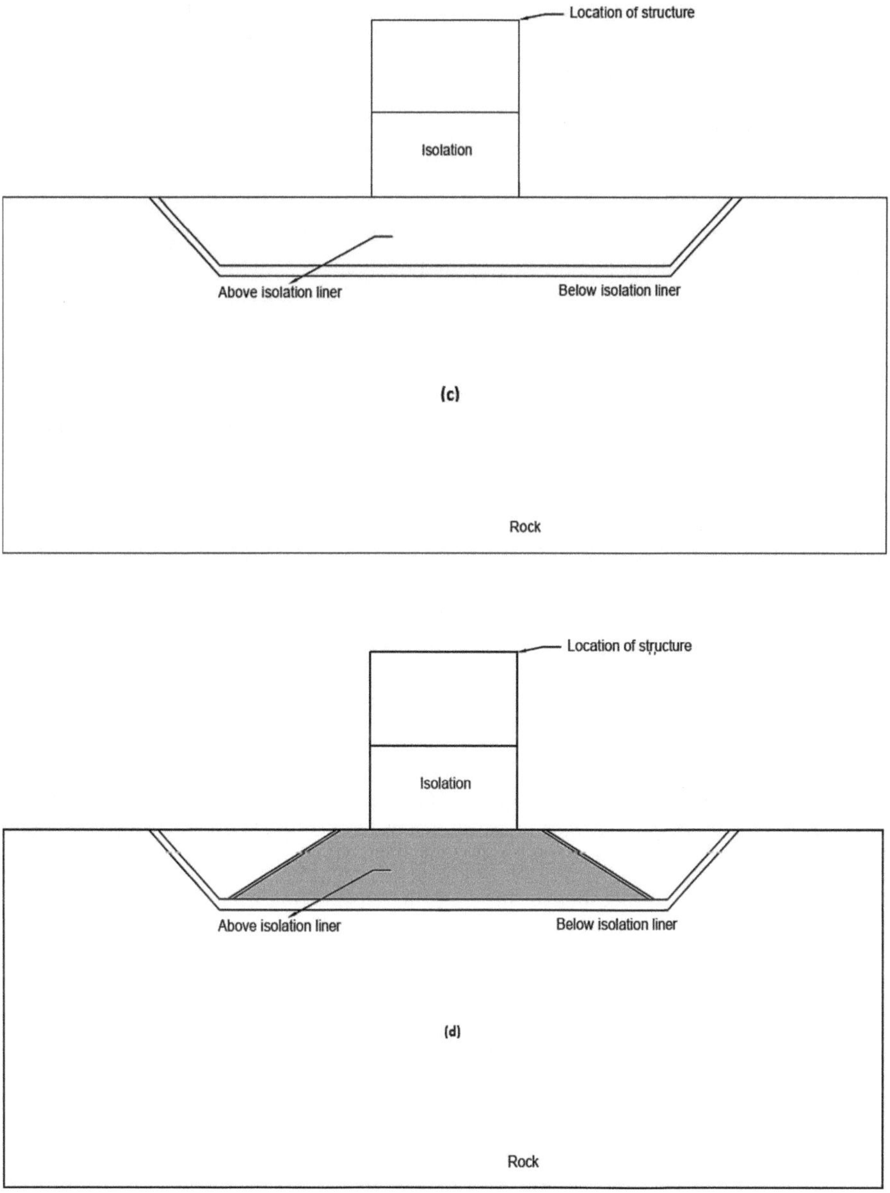

Fig. 1. (*continued*)

4 Studies on Natural Geotextiles Under Static Loading

Abedin et al. (1997) performed bearing capacity tests using square, strip and circular footings, on a sand bed in which jute cloth was placed at an intermediate depth within the homogenous sand. At smaller upper layer thicknesses, the soil mass above the reinforcing layer will have insufficient overburden to produce enough frictional resistance at an interface of the jute and the soil. The limited number of data indicated that an increase in friction angle of sand decreases the shape factor (i.e., the ratio of the observed bearing capacity of the reference bearing capacity).

Vinod and Bhaskar (2010) observed that at small values of embedment depth (i.e., $z < 0.33B$), the reinforcing mechanisms will have effect up to a lesser extent. This is because of less overburden pressure on the geotextile layer. As the embedment depth increases and if it goes beyond 0.5B, the overburden pressure increases on the geotextile layer and the soil above the geotextile layer causes settlement as the thickness of soil will be higher at greater embedment depths. If the length of the coir geotextile exceeds 1.2 times the size of the model footing a significant increase in strength improvement can be observed. This improvement of strength will be due to the dominance of the passive resistance mechanism which prevents the lateral movement of the soil mass from within the geotextile cells. From the results it can be said that the influence of vertical spacing between reinforcement layers with respect to the width of the footing and are interdependent, and hence the effect of any one of these cannot be isolated.

5 Studies on Natural Geocells Under Static Loading

Hegde and Sitharam (2015) proved analytically and experimentally that the bamboo geocells and geogrids together provides higher bearing capacity than commercially available geocells and geogrids. The soil infilled bamboo cell acts as a beam and disperses the loads acting on them to the wider areas. The basal geogrid ceases the downward movement of soil and reduces the settlement of the bed. The lateral confinement, vertical dispersion and membrane effects are the three mechanisms which contributes in increasing the bearing capacity of soil bed. The Neoloy and Bamboo geocells reduces the surface heave due to their interlocking effect (Fig. 2).

Fig. 2. Bamboo geocell (Source: Hegde and Sitharam 2017)

Lal et al. (2017) studied that coir geocells performs well than coir geotextiles under static loading. This is due to the frictional resistance in coir geocells. The performance of coir geocell will not be affected with the degradation of geocell as the soil attains the required strength with time due to applied pressure. They further specified that the coir coated with cement or bitumen will be more durable than uncoated coir. Further increase in width and no. of layers of geocell does not show any effect on the bearing capacity. Some other new geocells made of Areca leaf, coir fibers from coconut husk, jute fibers and sisal leaf (Figs. 3, 4, 5 and 6).

Fig. 3. Areca geocell

Fig. 4. Coir Geocell

Fig. 5. Jute Geocell

Fig. 6. Sisal Geocell

6 Studies on Geosynthetics Under Cyclic Loading

Vantamuri (2015) performed cyclic plate load tests on circular footing placed on geogrid reinforced sand bed prepared with different densities and with number of geogrid layers. Dynamic soil properties as coefficient of elastic uniform compression Cu, coefficient of elastic uniform shear $C\tau$, coefficient of elastic non-uniform shear $C\psi$ and the coefficient of elastic non uniform compression $C\varphi$ were determined. After performing several cyclic plate load tests, some key points were observed. As the density of sand bed and number of layers of geogrid increases, settlement of footing decreases substantially.

Sreedhar and Goud (2011) investigated the static and cyclic behavior of geosynthetic reinforced sand by conducting several cyclic load tests. Cyclic Plate Load tests were carried out separately on unreinforced sand and reinforced beds with different geosynthetic products. The results indicated that, the geosynthetic products with higher mobilized tensile strength and inter-surface friction will improve the bearing capacity of sand beds. But, the reasons behind this are not known.

Saran et al. (1995) proved that the damping capacity of the sand bed increases with the provision of geogrid reinforcement. After reinforcing sand bed with multiple layers of geogrids, the researchers observed huge changes in terms of settlement and bearing capacity of bed. As the reinforcement size and number increased, bearing capacity increased substantially which in return decreased the total settlement. The coefficient of elastic uniform compression reduces with the inclusion of reinforcing sheets. But this reduction continues up to certain bearing capacity. After this there will not be any change in the coefficient of elastic uniform compression.

7 Studies on Use of Polymer Geotextile Under Dynamic Loading

Yegian and Lahalf (1992) discussed geomembrane-geotextile and geomembrane-geomembrane interface systems under dynamic loading. It was observed that the limited shear force will be transmitted from a geomembrane to a geotextile or to another geomembrane. Whenever the lowest-interface friction force exceeds, permanent deformations will be accumulated along the interface of geosynthetic. After performing shake table tests, it was observed that the block acceleration was smaller than table acceleration, due to the relative displacement occurred along the geomembrane interface.

Yegian et al. (1995) conducted shake table tests and studied seismic response of geomembrane/geotextile and geomembrane/soil systems. The geosynthetic interface, whose shear strength is lesser than the surrounding soil, absorbs the energy through slip. It was observed that the slip deformations were occurred between the geotextile and the geomembrane when the table peak acceleration was greater than 0.2 g. The slip which occurs under a single pulse can be larger than the permanent slip. At the time of reversal of direction of motion, the shear force transmitted through the surface will increase momentarily.

Lahlaf and Yegian (1993) investigated the dynamic interface shear strength properties of geosynthetic systems using shaking table tests. The shake table tests provided values for both dynamic and static interface friction angles. Geomembrane is very sensitive to surface conditioning and wetting of geomembrane slightly reduces the dynamic friction angle of about 1 degree. Whereas the normal stress, the frequency of motion, the orientation of the geomembrane with respect to the direction of motion of the shaking table did not affect the dynamic friction angles. During shake table tests, the block initially moved with the table, up to a limiting table acceleration of about 0.2 g. The acceleration of the structure invariably drops, immediately after sliding of the block starts.

Georgarakos et al. (2005) discussed four in-ground isolation systems. The isolated soil wedge of cylindrical shape having circular sector of 60 m width and 5 m deep at middle is one of the in-ground isolation systems. This shape is expensive and difficult to achieve at the site than other geometries like tube liner, trapezoidal and compound trapezoidal (Fig. 1). Tube liner geometry which was 50 m long, 5 m deep and having two circular arcs of 5 m radius at the two edges showed higher peak acceleration on the surface when compared with cylindrical geometry. For this geometry the slippage at the liner interface was lower. This was mainly due to the side edges, which prevented the isolated soil from sliding freely on the horizontal plane. In this in-ground isolation system, plastic cracks were formed and the extent of these cracks was depended on soil strength characteristics of both isolated and the surrounding soil. In trapezoidal geometry (side angle - 30 or 60°), as the inclination of edges was milder, the acceleration on the isolated soil decreases and the slippage increases due to the decrease of the lateral opposing thrust from the surrounding soil. In compound trapezoidal liner geometry (side angle – 60°, double trapezium wedges), the loose-soil segments at both sides of the main core were forced to move up and down with some internal deformation, supplying the system with restoring force. The failure of this in-ground isolation system was unlikely.

Nanda et al. (2010) proposed that geotextile – smooth marble sliding interface can be effectively used to reduce base shear and roof acceleration. They conducted shake table tests on a scale model of 1999 mm × 1999 mm plan area with a sliding surface at plinth level and observed that the reduction of about 65% in the maximum acceleration response at the roof level of a building is possible. The reduction in absolute acceleration response occurred at the cost of relative displacement between the superstructure and substructure at the plinth level, if the plinth projection is within 75 mm.

Kavazanjian et al. (1991) conducted shake table tests and dynamic centrifuge model tests on a prototype model by placing a single synthetic liner beneath the base of the foundation. After conducting tests, they observed that the intensity of motions transmitted to the block had significantly reduced. From this they studied that if the interaction between the structure supported on a frictional base isolation and its foundation will further increase the effectiveness of the proposed system. Increase in sliding resistance was due to the boundary effects in the centrifuge and friction during the non-planar sliding block motions.

8 Conclusions and Recommendations

This paper reviewed several studies on geotextiles, geogrids and geocells (both synthetic and natural) under static and dynamic loading. It is observed that geocell perform better than geogrids due to the cellular confinement effect. The sufficient strength and stiffness of HDPE geocell provide long-term performance of a structure with smaller displacements. During earthquakes, ground movement will be in up and down and side to side. A two- dimensional geogrid can stabilize side to side earth movement, but it resists up and down movement on a small scale. A three- dimensional geocell which is infilled with soil or gravel will have both shear and bending moment resistance. This makes the geocell reinforced bed to exhibit a higher capacity to absorb and dissipate energy from side to side and up-down motion as well. The reinforcing geogrid and geocell layers also helps in resisting the rotational and translational failure.

However there are no studies available on natural fibers under dynamic loading. In addition to HDPE geocells, geocells made out of natural materials have great scope for potential applications in ground engineering. Figures show typical geocells made of natural material such as bamboo, areca leaf sheath, coir fiber, jute fiber and sisal fiber.

References

Rajagopal, K., Krishnaswamy, N.R., Madhavi, Latha G.: Behaviour of sand confined with single and multiple geocells. Geotext. Geomembr. **17**, 171–184 (1999). https://doi.org/10.1016/S0266-1144(98)00034-X. Elsevier

Chen, R.-H., Huang, Y.-W., Huang, F.-C.: Confinement effect of geocells on sand samples under triaxial compression. Geotext. Geomembr. **37**(2013), 35–44 (2013). https://doi.org/10.1016/j.geotexmem.2013.01.004. Elsevier

Pokharel, S.K., Han, J., Leshchinsky, D., Parsons, R.L., Halahmi, I.: Investigation of factors influencing behavior of single geocell-reinforced bases under static loading. Geotext. Geomembr. **28**(2010), 570–578 (2010). https://doi.org/10.1016/j.geotexmem.2010.06.002. Elsevier

Kargar, M., Mohammad Hosseini, S.M.: Effect of reinforcement geometry on the performance of a reduced-scale strip footing model supported on geocell-reinforced sand. Trans. A: Civ. Eng., Sci. Iran. A **24**(1), 96–109 (2017)

Moghaddas Tafreshia, S.N., Sharifia, P., Dawsonb, A.R.: Performance of circular footings on sand by use of multiple-geocell or -planar geotextile reinforcing layers. Soils Found. **56** (2016), 984–997 (2016). https://doi.org/10.1016/j.sandf.2012.01.002. Elsevier

Dash, S.K. Sireesh, G., Sitharam, T.G.: Model studies on circular footing supported on geocell reinforced sand underlain by soft clay. Geotext. Geomembr. **21**(2003), 197–219 (2003). https://doi.org/10.1016/S0266-1144(03)00017-7. Elsevier

Abedin, Z., Hasan, M., Dswan, S.A.: Bearing capacity of a jute cloth reinforced composite sand bed. In: Proceedings of the 14th International Conference on Soil Mechanics and Foundation Engineering. Humbarg, vol. 3, pp. 1553–1556 (2005)

Vinod, P., Bhaskar, Ajitha B.: Model studies on woven coir geotextile-reinforced sand bed. Proc. Inst. Civ. Eng.-Ground Improv. **165**(1), 53–57 (2010). https://doi.org/10.1680/grim.9.00030

Hegde, A., Sitharam, T.G.: Experimental and analytical studies on soft clay beds reinforced with bamboo cells and geocells. Int. J. Geosynth. Ground Eng. **1**, 1–13 (2015). https://doi.org/10.1007/s40891-015-0015-5

Lal, D., Sankar, N., Chandrakaran, S.: Effect of reinforcement form on the behaviour of coir geotextile reinforced sand beds. Soils Found. **57**(2017), 227–236 (2017). https://doi.org/10.1016/j.sandf.2016.12.001. Elsevier

Vantamuri, S.N.: Performance of cyclic loading on circular footing. Int. J. Res. Eng. Technol. **4** (5), 515–519 (2015)

Sreedhar, M.V.S., Goud, A.P.K.: Behaviour of Geosynthetic Reinforced Sand Bed Under Cyclic Load. In: Proceedings of Indian Geotechnical Conference, 15–17 December 2011, Kochi (Paper No.J-049) (2011)

Saran, S., Lavania, B.V.K., Sharma, R.K.: Cycle plate load tests on reinforced sand. In: Proceedings of Third International Conference on Recent Advances in Geotechnical Earthquake Engineering and Soil Dynamics, 2–7 April 1995, vol. 3, St. Louis, Missouri (1995)

Yegian, M.K., Lahalf, A.M.: Geomembranes as base isolation. Geosynthetic Fabric Report (1992)

Yegian, M.K., Yee, Z.Y., Harb, J.N.: Response of geosynthetics under earthquake excitations. In: Proceedings of Geosynthetics 1995 Conference, Nashville, TN, February, pp. 677–689 (1995)

Lahlaf, A.M., Yegian, M.K.: Shaking table tests for geosynthetic interfaces. In: Proceedings of the Geosynthetics 1993 Conference, Vancouver, Canada, March (1993)

Georgarakos, P., Yegian, M.K., Gazetas, G.: In-ground isolation using geosynthetic liners. In: Proceedings of 9th World Seminar on Seismic Isolation, Energy Dissipation and Active Vibration Control of Structures, Kobe, Japan, June (2005)

Nanda, R.P., Agarwal, P., Shrikhande, M.: Friction base isolation by geotextiles for brick masonry buildings. Geosynth. Int. **17**(1), 48–54 (2010). https://doi.org/10.1680/gein.2010.17.1.48

Kavanjian, E., Hushmand, B., Martin, G.R.: Frictional base isolation using a layered soil-synthetic liner system. ASCE: 1140-1151 (1991)

Pseudo-Dynamic Approach to Analyse the Effect of Soil Amplification in the Calculation of Seismic Active Earth Pressure Distribution Behind the Inclined Retaining Wall for Inclined c-ϕ Soil Backfill

Ashish Gupta(✉) and Vishwas A. Sawant

Department of Civil Engineering, IIT Roorkee, Roorkee, India
shi_g2000@rediffmail.com, sawntfce@iitr.ac.in

Abstract. In the current practice for the design of a retaining wall in the earthquake-prone regions, the study of seismic earth pressure is essential for the soil life-support system. Regarding the design of retaining walls in the earthquake-prone region, pseudo-static and pseudo-dynamic approaches are widely used for c-ϕ soil backfill without taking the effect of soil amplification. Soil amplification is very important and necessary factor to calculate the seismic active earth pressure analyzing the retaining walls. It should not be ignored in the design in the earthquake-prone regions. In this paper, a detailed formulation has been reported to calculate the seismic active earth pressure distribution along with the calculations of seismic active thrust for the inclined retaining wall. The retaining wall supports the inclined c-ϕ soil backfill. A parametric study is conducted to study the effect of various parameters like c-ϕ values of soil backfill, wall friction, soil amplification, horizontal and vertical seismic coefficients and the effect of time and phase difference in the shear waves and the primary waves. The results obtained for seismic earth pressure distribution is clearly showing the non-linear behavior behind the inclined retaining wall, which is the need for the design of retaining wall in earthquake-prone regions. The negative value of seismic active earth pressure distribution up to a normalized depth of wall is showing the zone of tension crack for the case of cohesive soil backfill.

List of notations

γ	Unit weight of soil (kN/m^3)
ω	Angular frequency (rad/s)
α	Assumed failure plane making an angle with the horizontal (°)
ψ	Time for propagating primary wave (s)
ζ	Time for propagating shear wave (s)
η	Wavelength of the vertically propagating primary wave (m)
a_f	Adhesion factor
a_h	Horizontal seismic acceleration
a_v	Vertical seismic acceleration
c	Soil cohesion (kN/m^2)
C	Total cohesive force (kN/m)

D. Choudhury et al. (Eds.): GeoMEast 2018, SUCI, pp. 87–100, 2019.
https://doi.org/10.1007/978-3-030-01920-4_8

c_a	Soil-wall adhesion (kN/m^2)
C_a	Total adhesive force (kN/m)
f_a	Soil amplification factor at top level
g	Acceleration due to gravity force
H	Height of wall (m)
i	Inclination of soil backfill (°)
K_{ae}(t)	Seismic active earth pressure coefficient
k_h	Horizontal seismic coefficient
k_v	Vertical seismic coefficient
P_{ae}(t)	Total seismic active thrust
p_{ae}(z,t)	Seismic active earth pressure distribution
$p_{ae}/\gamma H$	Seismic active earth pressure distribution (in non-dimensional form)
Q_h(t)	Total inertia force in horizontal direction (kN/m)
Q_v(t)	Total inertia force in vertical direction (kN/m)
T	Period of lateral shaking (s)
V_p	Primary wave velocity (m/s)
V_s	Shear wave velocity (m/s)
W	Weight of soil backfill (kN/m)
δ	Wall friction angle (°)
θ	Wall inclination with vertical (°)
λ	Wavelength of the vertically propagating shear wave (m)
ϕ	Soil friction angle (°)

1 Introduction

During the design of retaining wall in the earthquake prone regions, the critical study of seismic earth pressure is very crucial. To study the seismic active earth pressure, various researchers have been analyzed the retaining walls using the pseudo-static method. Using the pseudo-static method, the pioneer work for determining seismic earth pressure for the design of retaining walls had been reported by Okabe (1926) and Mononobe and Matsuo (1929) and then known as Mononobe and Okabe method. The time dependent effect during the earthquake loading was completely missing in the pseudo-static method. In pseudo-static method, magnitude and phase of seismic accelerations were also taken uniform throughout the soil backfill.

For analyzing the real seismic problems during the design of retaining walls, Steedman and Zeng (1990) had proposed the pseudo-dynamic approach considering the finite shear waves in the backfiill. Pseudo-dynamic method had been used to overcome the deficiencies of pseudo-static method. Choudhury and Nimbalkar (2005 and 2006) had extended that pseudo-dynamic approach for determining the seismic passive and seismic active earth pressure behind vertical retaining wall. Nimbalkar and Choudhury (2008) introduced the soil amplification effect in the pseudo-dynamic approach to determine the seismic active earth pressure coefficients and seismic active earth pressure distribution for vertical retaining wall. Ghosh (2008) had extended the study of Nimbalkar and Choudhury (2008) for the inclined retaining wall. Gupta and Sawant (2018) studied the effect of soil amplification on the dynamic response of retaining wall having cohesionless backfill. Both wall and backfill are having nonzero inclination with vertical and horizontal. Using the pseudo-dynamic approach considering c-ϕ soil backfill, Ghosh and Sharma (2010) had reported the effect of tension crack in the top portion of the backfill. The tension crack was determined by the Rankine's analysis of active earth pressure for c-ϕ soil backfill under static case, which was the limitation of this study. Shao-jun et al. (2012) extended the study of Ghosh and Sharma (2010) to investigate the effect of tension crack under seismic loading in the top portion of the backfill, without considering the effect of soil modification factor.

In the present work, a detailed formulation incorporating the soil amplification effect has been reported to compute the seismic active earth pressure distribution. The retaining wall is inclined supporting the inclined c-ϕ soil backfill.

2 Detailed Formulation

The rigid inclined retaining wall AB of height H inclined at an angle θ with vertical and wall friction angle δ as shown in Fig. 1. It is retaining soil backfill of cohesion c and soil friction angle ϕ, having unit weight γ inclined at an angle i with horizontal. Effect of propagation of both shear waves and the primary waves is also considered along with the effect of soil amplification. Linear variation in input ground acceleration along depth is taken for showing the effect of soil amplification, within the soil media due to the seismic loading. Amplitude of horizontal and vertical seismic acceleration of base of the retaining wall assumed as $a_h = k_h$ g and $a_v = k_v$ g. The primary and shear wave velocities, V_s and V_p are assumed to propagate within the soil media due to the earthquake.

The horizontal and vertical seismic acceleration at the top has been assumed, higher than the value of horizontal and vertical seismic acceleration at the base. In the present work, horizontal and vertical seismic acceleration at the top of the retaining wall is taken as $k_{h(z=0)} = f_a.k_{h(z=H)}$ and $k_{v(z=0)} = f_a.k_{v(z=H)}$, where f_a is the soil amplification factor. The present analysis induces a period of lateral shaking $T = 2\pi/\omega$, where ω is the angular frequency.

A failure wedge ABE is assumed, which makes an angle α with horizontal. W is the weight of failure wedge. The horizontal and vertical inertia forces are $Q_h(t)$ and $Q_v(t)$. C is the total cohesive force and C_a is the total soil-wall adhesion force. Soil-wall adhesion factor is taken as a_f which defines $a_f = (C_a/C) = (c_a/c)$, where c_a is the soil-wall adhesion.

From Fig. (1), the mass of the strip of thickness dz at depth z can be obtained as:

$$m(z) = \frac{\gamma}{g} m(\alpha)(H - z)dz \tag{1a}$$

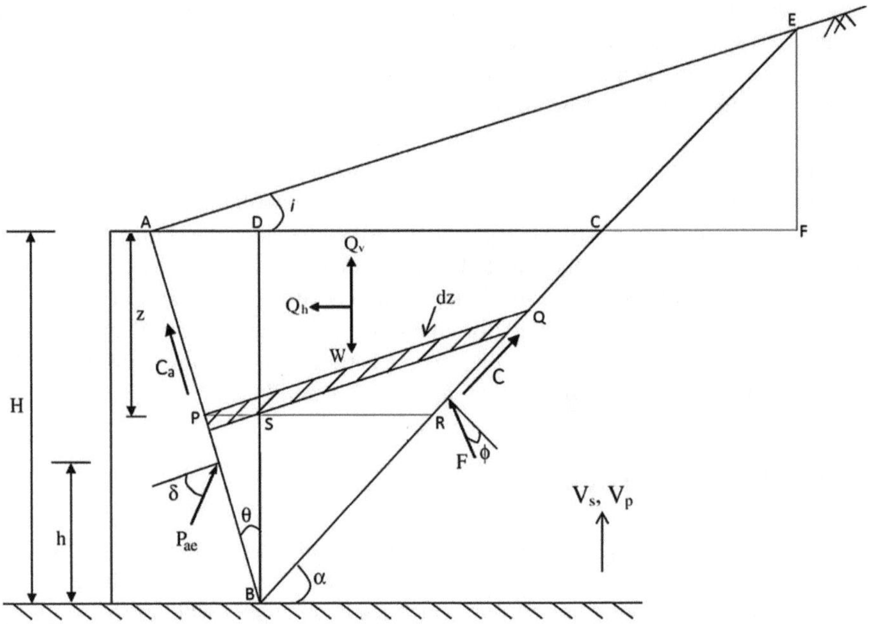

Fig. 1. Forces acting on retaining wall in active state

where,

$$m(\alpha) = \frac{(1 + \tan\alpha \tan\theta)\ \sin\alpha\ \cos(\theta - i)}{\tan\alpha\ \sin(\alpha - i)\ \cos\theta} \tag{1b}$$

The weight of the failure wedge can be obtained as:

$$W = \frac{1}{2}\gamma H^2 m(\alpha) \tag{2}$$

At any depth z below the top of the wall, the horizontal and the vertical acceleration can be expressed as:

$$a_h(z,t) = \left\{1 + \frac{H-z}{H}(f_a - 1)\right\} k_h\ g\ \sin\omega\left(t - \frac{H-z}{V_S}\right) \tag{3a}$$

$$a_v(z,t) = \left\{1 + \frac{H-z}{H}(f_a - 1)\right\} k_v\ g\ \sin\omega\left(t - \frac{H-z}{V_P}\right) \tag{3b}$$

The total horizontal inertia force $Q_h(t)$ acting in the failure wedge is given by:

$$Q_h(t) = \int_0^H m(z)\, a_h(z,t) = \gamma\, k_h\, m(\alpha)\, (I_1 + I_2) \tag{4}$$

where,

$$I_1 = \frac{\lambda}{4\pi^2}[2\pi H\, \cos\omega\zeta + \lambda(\sin\omega\zeta - \sin\omega t)] \tag{5a}$$

$$I_2 = \frac{(f_a - 1)\lambda}{4\pi^3 H}\{2\pi H(\pi H\, \cos\omega\zeta + \lambda\, \sin\omega\zeta) + \lambda^2(\cos\omega t - \cos\omega\zeta)\} \tag{5b}$$

The total horizontal inertia force $Q_h(t)$ can be rewritten as:

$$Q_h(t) = \frac{\lambda\gamma k_h m(\alpha)}{4\pi^2}\left[\left\{\begin{matrix} 2\pi H\, \cos\omega\zeta \\ +\lambda(\sin\omega\zeta - \sin\omega t) \end{matrix}\right\} + \frac{(f_a - 1)}{\pi H}\left\{\begin{matrix} 2\pi H\begin{pmatrix} \pi H\, \cos\omega\zeta \\ +\lambda\, \sin\omega\zeta \end{pmatrix} \\ +\lambda^2(\cos\omega t - \cos\omega\zeta) \end{matrix}\right\}\right] \tag{6}$$

Using the same procedure for calculating the total vertical inertia force $Q_v(t)$ acting in the failure wedge is given by:

$$Q_v(t) = \frac{\eta\gamma k_v m(\alpha)}{4\pi^2}\left[\left\{\begin{matrix} 2\pi H\, \cos\omega\psi \\ +\eta(\sin\omega\psi - \sin\omega t) \end{matrix}\right\} + \frac{(f_a - 1)}{\pi H}\left\{\begin{matrix} 2\pi H\begin{pmatrix} \pi H\, \cos\omega\psi \\ +\eta\, \sin\omega\psi \end{pmatrix} \\ +\eta^2(\cos\omega t - \cos\omega\psi) \end{matrix}\right\}\right] \tag{7}$$

where,

$$\lambda = TV_S; \eta = TV_P; \zeta = \left(t - \frac{H}{V_S}\right) \text{and } \psi = \left(t - \frac{H}{V_P}\right)$$

Resolving the forces in the horizontal and vertical direction on the failure wedge, the total active thrust $P_{ae}(t)$ can be obtained as:

$$P_{ae}(t) = \left[\begin{matrix} \left\{\dfrac{\{W - Q_v(t)\}\sin(\alpha-\phi) + Q_h(t)\cos(\alpha-\phi)}{\cos(\delta+\theta-\alpha+\phi)}\right\} \\ -\left\{\dfrac{cH\{\cos(\theta-i)\cos\phi - a_f\sin(\theta-\alpha+\phi)\sin(\alpha-i)\}}{\sin(\alpha-i)\cos\theta\,\cos(\delta+\theta-\alpha+\phi)}\right\} \end{matrix}\right] \tag{8}$$

Using Eq. (8), the seismic active earth pressure coefficient, $K_{ae}(t)$ can be obtained as:

$$K_{ae}(t) = \begin{bmatrix} \left\{ \frac{2\{W - Q_v(t)\}\sin(\alpha-\phi) + 2Q_h(t)\cos(\alpha-\phi)}{\gamma H^2 \cos(\delta+\theta-\alpha+\phi)} \right\} \\ -\frac{2c}{\gamma H}\left\{ \frac{\{\cos(\theta-i)\cos\phi - a_f \sin(\theta-\alpha+\phi)\sin(\alpha-i)\}}{\sin(\alpha-i)\cos\theta\cos(\delta+\theta-\alpha+\phi)} \right\} \end{bmatrix} \quad (9)$$

Substituting the values of W, Q_h(t) and Q_v(t) in Eq. (9), an expression for K_{ae}(t) can be derived. On optimizing Eq. (9) with respect to their two variables α and t/T, it can be obtain the maximum value of K_{ae}(t).

On taking the partial derivative of P_{ae}(t) with respect to z, seismic active earth pressure distribution behind the retaining wall can be determined and expressed as:

$$p_{ae}(z,t) = \frac{\partial P_{ae}(z,t)}{\partial z} = \{D_1 + D_2 + D_3 + D_4 + D_5 + D_6\} \quad (10)$$

where,

$$D_1 = m(\alpha)\frac{\gamma\, z\, \sin(\alpha-\phi)}{\cos(\delta+\theta-\alpha+\phi)} \quad (11a)$$

$$D_2 = m(\alpha)\frac{k_h\, \gamma\, z\, \cos(\alpha-\phi)}{\cos(\delta+\theta-\alpha+\phi)}\sin\theta_{\lambda z} \quad (11b)$$

$$D_4 = -m(\alpha)\frac{k_v\, \gamma\, z\, \sin(\alpha-\phi)}{\cos(\delta+\theta-\alpha+\phi)}\sin\theta_{\eta z} \quad (11c)$$

$$D_3 = \begin{bmatrix} \left\{ m(\alpha)\frac{k_h\,\gamma\,(f_a-1)\cos(\alpha-\phi)}{\cos(\delta+\theta-\alpha+\phi)}\frac{\lambda}{2\pi} \right\} \\ \left\{ -\cos\theta_{\lambda z} - \frac{\lambda}{\pi z}\sin\theta_{\lambda z} + \frac{\lambda^2}{2\pi^2 z^2}\{\cos\theta_{\lambda z} - \cos\theta_t\} + \frac{2\pi z}{\lambda}\sin\theta_{\lambda z} \right\} \end{bmatrix} \quad (11d)$$

$$D_5 = -\begin{bmatrix} \left\{ m(\alpha)\frac{k_v\,\gamma\,(f_a-1)\sin(\alpha-\phi)}{\tan\alpha\,\cos(\delta+\theta-\alpha+\phi)}\frac{\eta}{2\pi} \right\} \\ \left\{ -\cos\theta_{\eta z} - \frac{\eta}{\pi z}\sin\theta_{\eta z} + \frac{\eta^2}{2\pi^2 z^2}\{\cos\theta_{\eta z} - \cos\theta_t\} + \frac{2\pi z}{\eta}\sin\theta_{\eta z} \right\} \end{bmatrix} \quad (11e)$$

$$D_6 = -\frac{c\{\cos(\theta-i)\cos\phi - a_f\sin(\theta-\alpha+\phi)\sin(\alpha-i)\}}{\sin(\alpha-i)\cos\theta\cos(\delta+\theta-\alpha+\phi)} \quad (11f)$$

$$\theta_{\lambda z} = 2\pi\left(\frac{t}{T} - \frac{z}{\lambda}\right) \text{ and } \theta_{\eta z} = 2\pi\left(\frac{t}{T} - \frac{z}{\eta}\right) \quad (11g)$$

By optimizing Eq. (9) with respect to α and t/T, optimized values of α and t/T are obtained. Using these two optimized values, terms D_1, D_2, D_3, D_4, D_5 and D_6 are evaluated. On putting D_1, D_2, D_3, D_4, D_5 and D_6 in Eq. (10), it provides a general expression of seismic active earth pressure distribution behind the inclined wall considering inclined c-ϕ soil backfill.

3 Results and Discussion

Seismic active earth pressure distribution is presented in the non-dimensional ($p_{ae}/\gamma H$) using Eq. (10). In the present study, a parametric study has been done to quantify the value of $p_{ae}/\gamma H$ along the normalized depth of the retaining wall. Variation of parameters considered are stated in Table 1.

Table 1. Variation of parameters considered in the present study

S. No.	Description	Values are taken
1.	Unit weight of soil backfill (γ)	20 kN/m^3
2.	The height of retaining wall (H)	10 m
3.	Shear wave velocity (V_S)	100 m/s
4.	Primary wave velocity (V_P)	187 m/s
5.	The time period of lateral shaking (T)	0.3 s
6.	Soil cohesion (c)	0, 10 and 20 kPa
7.	Soil friction angle (ϕ)	25°, 30°, 40° and 50°
8.	Wall inclination with vertical (θ)	−30°, −15°, 0°, 15° and 30°
9.	The inclination of soil backfill with horizontal (i)	0° and 10°
10.	Wall friction angle (δ)	0.5 ϕ
11.	Coefficient of horizontal seismic acceleration (k_h)	0.0, 0.1, 0.2, and 0.3
12.	Coefficient of vertical seismic acceleration (k_v)	0.0, 0.25 k_h, 0.5 k_h and 0.75 k_h
13.	Soil amplification factor (f_a)	1.0, 1.2, 1.4, 1.6 and 1.8

3.1 Effect of f_a on $p_{ae}/\gamma H$

Figures 2, 3, 4 and 5 are showing the variation of $p_{ae}/\gamma H$ along the normalized depth (z/H) for the soil amplification factor (f_a) varying from 1.0 to 1.8. On increasing the value of f_a from 1.0 to 1.8, it can be clearly observed that the value of $p_{ae}/\gamma H$ increases significantly. From the Figs. 2, 3, 4 and 5 for the higher value of f_a (for $z/H \geq 0.5$), the effect of f_a is significantly decreases for the case of c-ϕ soil backfill. From Figs. 2, 3, 4 and 5, it can be also observed that the value of $p_{ae}/\gamma H$ increases considerably faster when f_a increases from 1.4 to 1.8 as compare to f_a increases from 1.0 to 1.4 for four cases ($c = 0$ and $a_f = 0.0$; $c = 10$ kPa and $a_f = 0.0$; $c = 10$ kPa and $a_f = 1.0$; $c = 20$ kPa and $a_f = 1.0$). For example, at $z/H = 0.8$ (for $c = 0$ and $a_f = 0.0$), $p_{ae}/\gamma H$ increases about 11% when f_a increases from 1.0 to 1.4 and 14% when f_a increases from 1.4 to 1.8 (for $i = 0°$) and about 20.3% and 34.7% respectively for $i = 10°$. The same is increases about 10.1% and 11.9% (for $i = 0°$) and about 15.5% and 20.6% (for $i = 10°$) for $c = 10$ kPa and $a_f = 0.0$. The value of $p_{ae}/\gamma H$ is increases about 11.3% and 13.3% (for $i = 0°$) and about 16.7% and 21.8% (for $i = 10°$) for $c = 10$ kPa and $a_f = 1.0$; which shows a little bit effect of soil-wall adhesion. On increasing the value of c as 20 kPa keeping $a_f = 1.0$; $p_{ae}/\gamma H$ is increases about 11.8% and 13.3% (for $i = 0°$) and about 15.3% and 18.1% (for $i = 10°$). The negative value of $p_{ae}/\gamma H$ in all cases is showing the zone of tension crack due to the cohesive soil backfill. For $i = 10°$, on increasing the value of f_a as 1.0, 1.2, 1.4, 1.6 and 1.8, the value of depth of tension crack increases respectively as 1.01 m, 1.095 m, 1.215 m, 1.385 m and 1.595 m (for c = 10 kPa and a_f = 1.0). For the same case, the depth of tension crack increases as 1.736 m, 1.781 m, 1.842 m, 1.919 m and 2.017 m (for c = 20 kPa and a_f = 1.0).

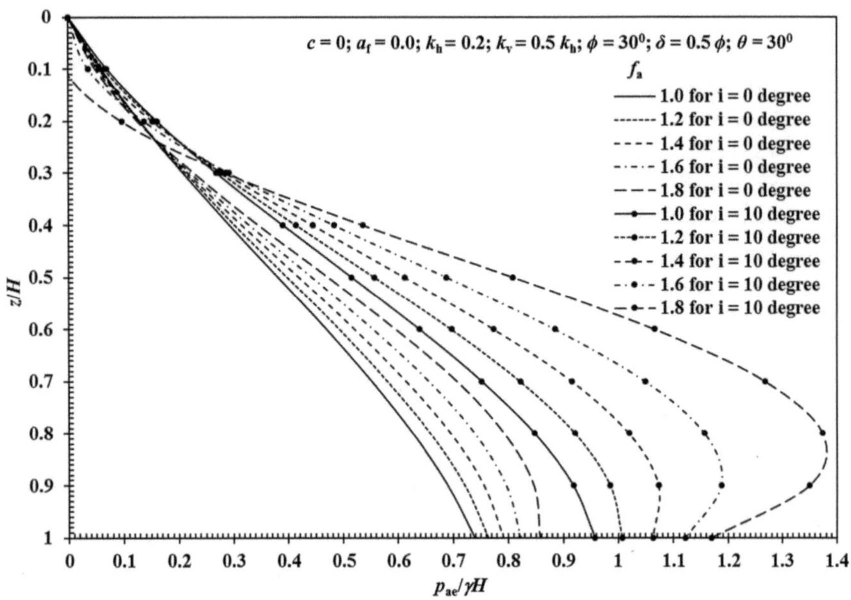

Fig. 2. ($p_{ae}/\gamma H$) with (z/H) for different values of f_a (for c = 0, a_f = 0.0)

3.2 Effect of k_h and k_v on $p_{ae}/\gamma H$

The variation of $p_{ae}/\gamma H$ along the normalized depth (z/H) for different values of k_h, varying from 0.0 to 0.3 is shown in Fig. 6. The effect of k_v on the value of $p_{ae}/\gamma H$ is also shown in Fig. 7. Effect of soil amplification is also taken in the variation of $p_{ae}/\gamma H$ for both k_h and k_v. It can be noticed from Fig. 6, that the value of $p_{ae}/\gamma H$ increases continuously, when the value of k_h increases from 0.0 to 0.3 (for $z/H > 0.3$). The same trend can be noticed when the value of k_v increases from $0.0k_h$ to $0.75k_h$ for $k_h = 0.3$ (for $z/H > 0.4$). On increasing the values of k_h, the behavior of $p_{ae}/\gamma H$ changes from linear to non-linear. The considerable increase of $p_{ae}/\gamma H$ can be also observed when k_h increases from 0.2 to 0.3 and k_v increases from $0.50k_h$ to $0.75k_h$ (for c = 10 kPa and a_f = 1.0; f_a = 1.4; i = 10°). For example, for z/H = 0.8, $p_{ae}/\gamma H$ increases about 68.3% when k_h increases from 0.0 to 0.2 and 81.3% when k_h increases from 0.2 to 0.3. For the same case $p_{ae}/\gamma H$ increases about 13.9% (k_v increases from $0.0k_h$ to $0.50k_h$) and 21.1% (k_v increases from $0.50k_h$ to $0.75k_h$). On increasing the value of k_h as 0.0, 0.1, 0.2 and 0.3, the value of depth of tension crack increases respectively as 0.86 m, 0.91 m, 1.215 m and 2.410 m for c = 10 kPa and a_f = 1.0 (for i = 10°). For the same case with

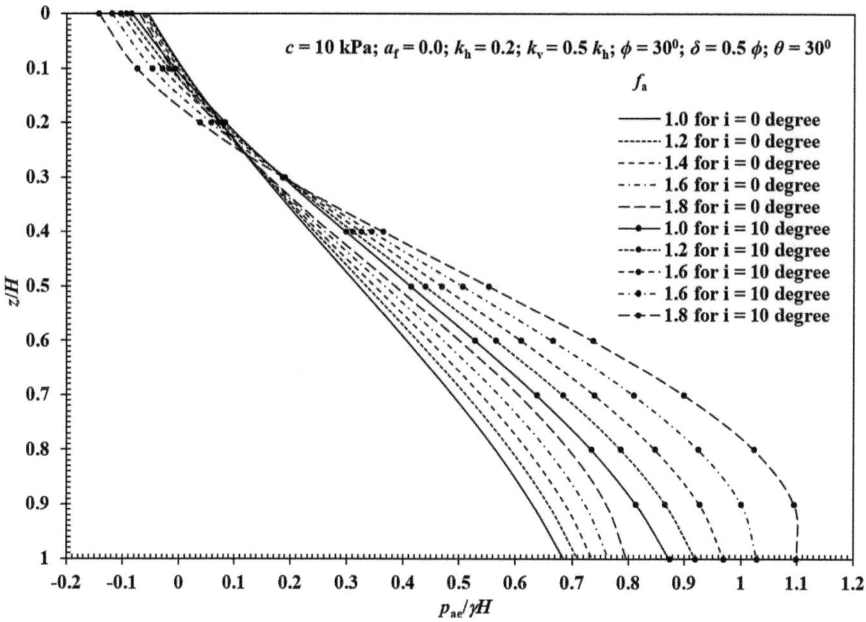

Fig. 3. (p_{ae}/γ *H*) with (*z*/*H*) for different values of f_a (for c = 10 kPa, a_f = 0.0)

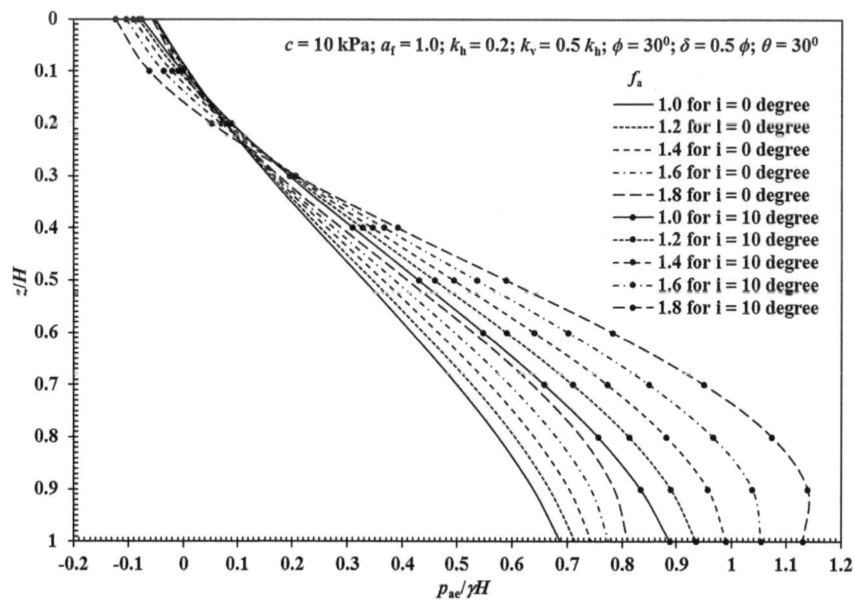

Fig. 4. (p_{ae}/γ *H*) with (*z*/*H*) for different values of f_a (for c = 10 kPa, a_f = 1.0)

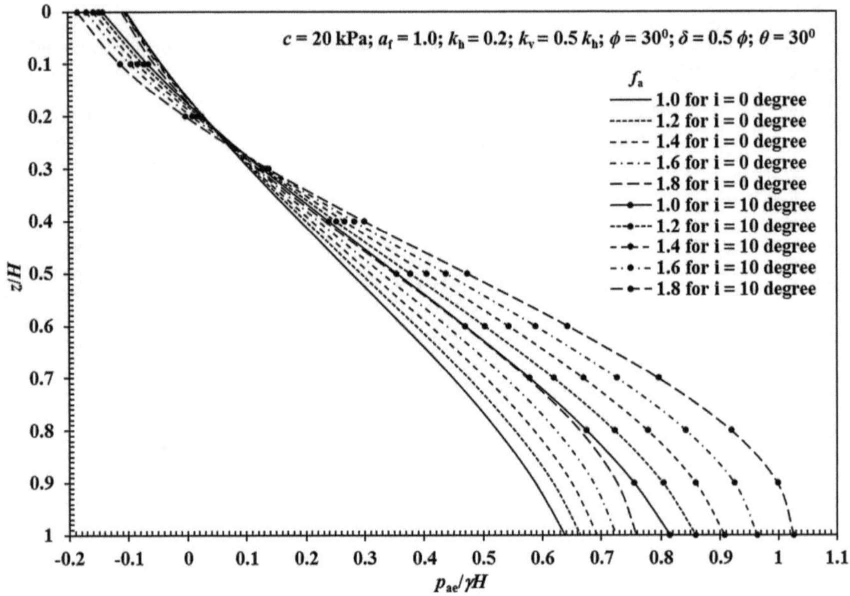

Fig. 5. ($p_{ae}/\gamma\ H$) with (z/H) for different values of f_a (for c = 20 kPa, a_f = 1.0)

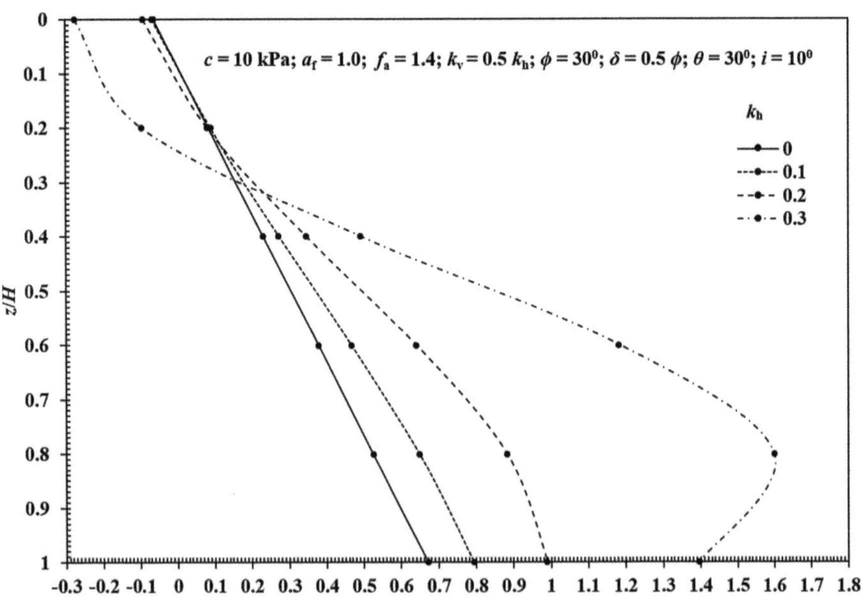

Fig. 6. ($p_{ae}/\gamma\ H$) with (z/H) for different values of k_h (for c = 10 kPa, a_f = 1.0)

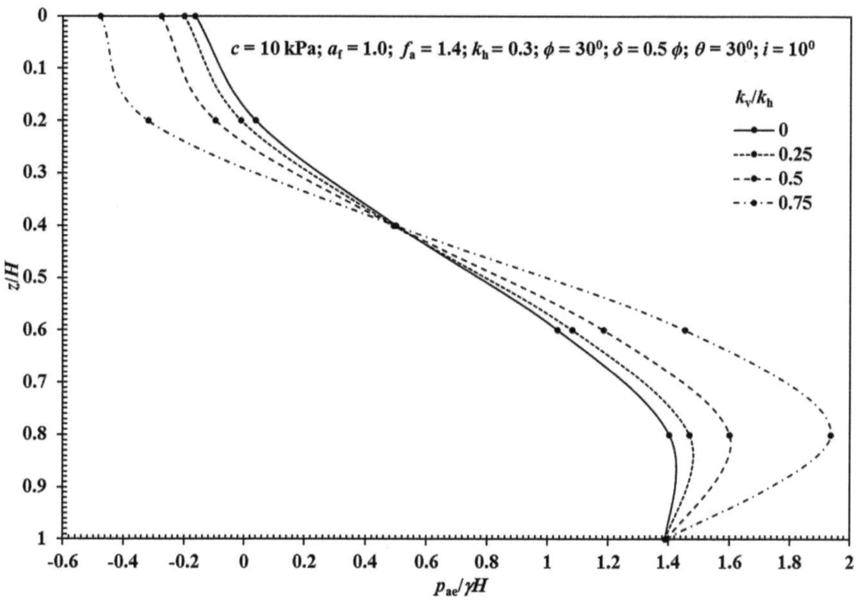

Fig. 7. ($p_{ae}/\gamma H$) with (z/H) for different values of k_v (for c = 10 kPa, a_f = 1.0)

the values of k_v as 0.0 k_h, 0.25k_h, 0.50k_h and 0.75k_h; the depth of tension crack increases as 1.771 m, 2.054 m, 2.412 m and 2.916 m.

3.3 Effect of θ on $p_{ae}/\gamma H$

The effect of wall inclination θ on $p_{ae}/\gamma H$ along z/H is shown in Fig. 8 for θ from −30°, −15°, 0°, 15° and 30° (for c = 10 kPa and a_f = 1.0; f_a = 1.2; i = 10°). As the wall moves from negative inclination to the positive inclination, the value of $p_{ae}/\gamma H$ increases more, for θ more than 0°. For example, for z/H = 0.8, $p_{ae}/\gamma H$ increases about 79.8%, and 125.2% when θ increases from −30° to 0° and 0° to 30°. On increasing the value of θ from −30°, −15°, 0°, 15° and 30°, the value of depth of tension crack increases respectively as 4.37 m, 3.00 m, 2.13 m, 1.52 m and 1.10 m for c = 10 kPa and a_f – 1.0 (for i – 10°).

3.4 *Effect of* ϕ on $p_{ae}/\gamma H$

The effect of soil friction angle ϕ on the value of $p_{ae}/\gamma H$ along the normalized depth (z/H) is shown in Fig. 9 for ϕ increases from 25° to 50° (for c = 10 kPa and a_f = 1.0; f_a = 1.2; i = 10°). The value of $p_{ae}/\gamma H$ decreases more, when ϕ increases from 30° to 40°. For example, for z/H = 0.8, $p_{ae}/\gamma H$ decreases about 12.5%, 18.8% and 15.4% when ϕ increases from 25° to 30°, 30° to 40° and 40° to 50°. On increasing the value of ϕ as 25°, 30°, 40° and 50°, the value of depth of tension crack decreases respectively as 1.27 m, 1.10 m, 0.94 m and 0.79 m for c = 10 kPa and a_f = 1.0 (for i = 10°).

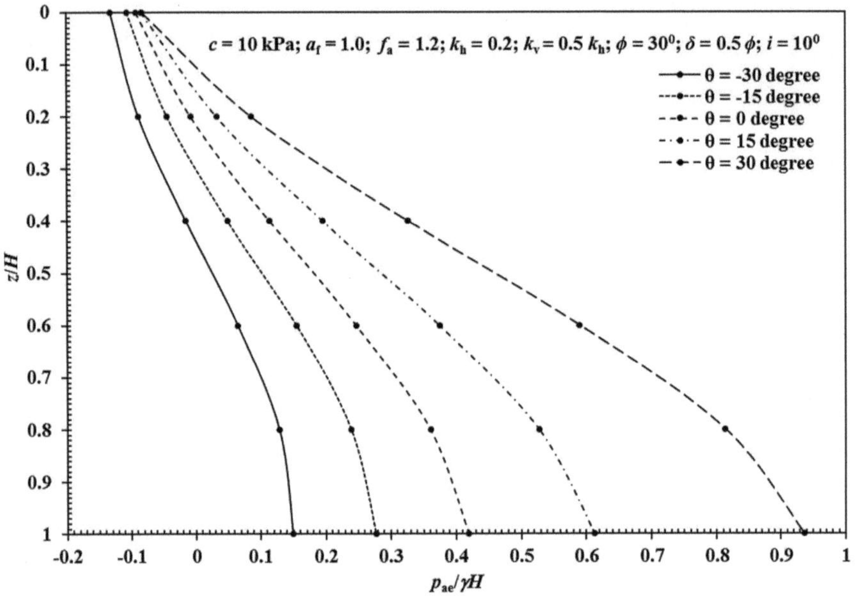

Fig. 8. ($p_{ae}/\gamma\ H$) with (z/H) for different values of θ (for c = 10 kPa, a_f = 1.0)

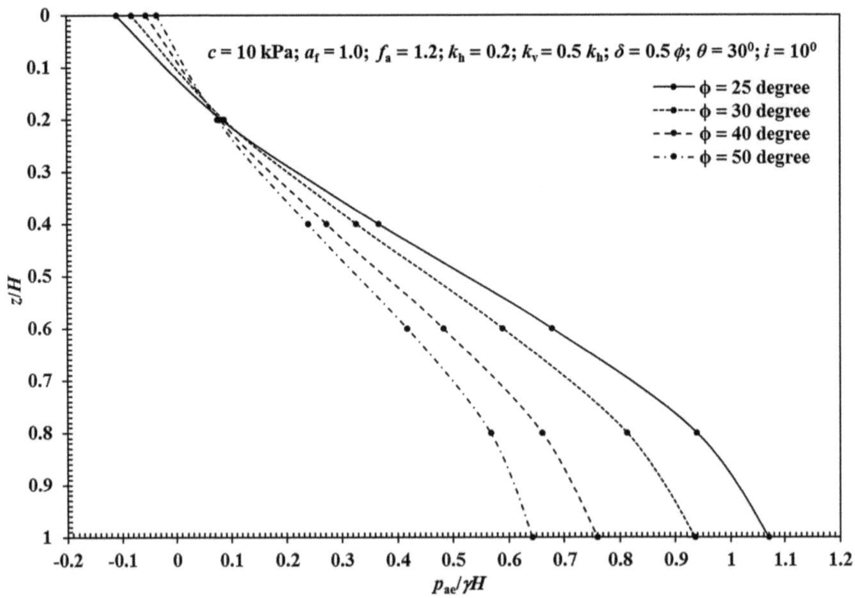

Fig. 9. ($p_{ae}/\gamma\ H$) with (z/H) for different values of ϕ (for c = 10 kPa, a_f = 1.0)

4 Comparision of Results

The depth of tension crack obtained in the present study have been compared with the values reported by Ghosh and Sharma (2010) and Shao-jun et al. (2012) for a set of parameters ($\theta = 10°$, $\delta = \phi/2$, $a_f = 0.5$, $k_h = 0.2$ and $k_v = 0.1$) which is shown in Table 2. The values of the depth of tension crack obtained from the present analysis are lower than those obtained by Ghosh and Sharma (2010) and Shao-jun et al. (2012) for the same set of parameters. The difference in the values is quite significant for the higher values of cohesion of soil backfill.

Table 2. Comparison of depth of tension crack obtained from present study with earlier analytical works (for $\theta = 10°$, $\delta = \phi/2$, $a_f = 0.5$, $k_h = 0.2$ and $k_v = 0.1$)

ϕ (°)	c (kPa)	Depth of tension crack		
		Present study	Shao-jun et al. (2012)	Ghosh and Sharma (2010)
20	10	0.147	0.178	0.199
40	10	0.173	0.210	0.298
20	20	0.268	0.325	0.396
40	20	0.315	0.381	0.596

5 Conclusions

The detailed formulation is reported for calculating the seismic earth pressure distribution behind the inclined retaining wall. These formulations are obtained for the inclined cohesive soil backfill including the effect of soil amplification. From the equation of seismic earth pressure distribution, it can be clearly observed the non-linear behavior behind the inclined retaining wall in the pseudo-dynamic analysis, which shows the actual behavior of retaining wall under seismic condition. The negative value of the seismic active earth pressure distribution shows the zone of tension cracks developed for the case of c-ϕ soil backfill under seismic condition. The main conclusions drawn from the present study are as follows:

1. Non-dimensional value of seismic active earth pressure distribution increases when soil amplification factor increases for increasing depth. The effect is more significant for soil amplification factor more than 1.4. For the case inclined retaining wall for the cohesionless soil backfill the non-linear behavior is more, which reduces for the case of cohesive soil backfill.
2. On increasing the values of horizontal seismic coefficient, non-dimensional seismic active earth pressure distribution increases substantially for the horizontal seismic coefficient value more than 0.1. Non-linear behavior is also increases for the higher value of horizontal seismic coefficient.
3. On increasing the wall inclination for its negative to positive values, non-dimensional seismic active earth pressure distribution increases very fast for the positive values of the wall inclination.

4. Non-dimensional value of seismic active earth pressure distribution decreases when soil friction angle increases for increasing depth. For soil friction angle more than 25° and less than 40°, the effect is more.
5. The depth of tension crack increases when soil amplification factor increases for increasing depth for the case of cohesive soil backfill. The depth of tension crack also increases, when the angle of shearing resistance increases.
6. The depth of tension crack is high for the negative value of the wall inclination, which decreases for the positive value of the wall inclination.

References

Choudhury, D., Nimbalkar, S.S.: Seismic passive resistance by Pseudo-dynamic method. Geotechnique **55**(9), 699–702 (2005)

Choudhury, D., Nimbalkar, S.S.: Pseudo-dynamic approach of seismic active earth pressure behind retaining wall. Geotech. Geol. Eng. **24**(5), 1103–1113 (2006). Springer

Ghosh, P.: Seismic active earth pressure behind a nonvertical retaining wall using pseudo-dynamic analysis. Can. Geotech. J. **45**, 117–123 (2008)

Ghosh, S., Sharma, R.P.: Pseudo-dynamic active response of non-vertical retaining wall supporting c-ϕ backfill. Geotech. Geol. Eng. **28**(5), 633–641 (2010). Springer

Gupta, A., Sawant, V.A.: Effect of soil amplification on seismic earth pressure using pseudo-dynamic approach. Int. J. Geotech. Eng. (2018). https://doi.org/10.1080/19386362.2018.1476803

Mononobe, N., Matsuo, H.: On the determination of earth pressure during earthquakes. In: Proceedings of 2nd World Engineering Conference, Tokyo, Japan, 9, paper no. 388, pp. 177–185 (1929)

Nimbalkar, S.S., Choudhury, D.: Effects of body waves and soil amplification on seismic earth pressure. J. Earthq. Tsunami **2**(1), 33–52 (2008)

Okabe, S.: General theory of earth pressure. J. Jpn. Soc. Civ. Eng. **12**(1), 1277–1323 (1926)

Shao-jun, M.A., Kui-hua, W., Wen-bing, W.U.: Pseudo-dynamic active earth pressure behind retaining wall for cohesive soil backfill. J. Cent. South Univ. **19**, 3298–3304 (2012). https://doi.org/10.1007/s11771-012-1407-5. Springer

Steedman, R.S., Zeng, X.: The influence of phase on the calculation of pseudo-static earth pressure on a retaining wall. Geotechnique **40**(1), 103–112 (1990)

Large Scale Geotechnical Shake Table Testing at the University of California San Diego

Ahmed Ebeido[1(✉)], Muhammad Zayed[1], Kyungtae Kim[1], Patrick Wilson[2], and Ahmed Elgamal[1]

[1] Department of Structural Engineering, University of California San Diego, San Diego, USA
{aebeido, mzayed}@eng.ucsd.edu, {kyk002, aelgamal}@ucsd.edu

[2] Earth Mechanics, Inc, San Marcos, CA, USA
p.wilson@earthmech.com

Abstract. This paper discusses a range of geotechnical testing capabilities available at the University of California, San Diego (UCSD). The capabilities discussed are part of overall large-scale UCSD laboratories, one of the largest and most active earthquake engineering laboratories worldwide. Specifically, testing using 2 different shake tables is discussed, one being the largest outdoor shake table in the world. Illustration of the capabilities are shown for soil-structure-interaction (SSI) tests of embedded structures as well as pile foundations in dry and saturated soil scenarios. The testing models were densely instrumented with top of the line sensor arrays. Instrumentation includes strain gauges, pressure and excess pore-pressure sensors, accelerometers and displacement pots. The method of construction is presented, and the stages clearly described.

1 Introduction

Physical modelling is a valuable resource to complement the field investigations as the observed quantities are scarce. As such, centrifuge testing emerged as an efficient and cost effective method to study any problem. However, having its limitations, such as material use and scaling laws (Ubilla et al. 2011). Large-scale tests provide an advantage of having densely instrumented models, instrumenting both sides of a structure heavily and the ability to use genuine unscaled material such as reinforced concrete. Combining the use of centrifuge and large-scale testing benefits the testing community. Centrifuge testing narrows the needed large-scale test range, thus reducing the number of the expensive and time consuming large-scale tests.

Large servo-hydraulic shaking table systems are vital tools in experimental earthquake engineering. They provide effective ways to subject structural components, substructures, or entire structural systems to dynamic excitations similar to those induced by real earthquakes (Ozcelik et al. 2008). University of California, San Diego, hence referred to as UCSD has 2 shake tables in its facilities. A medium sized shake table located in the campus (Powell) lab and a large sized shake table located 15 km off-campus at the Englekirk Structural Engineering Center.

D. Choudhury et al. (Eds.): GeoMEast 2018, SUCI, pp. 101–113, 2019.
https://doi.org/10.1007/978-3-030-01920-4_9

The on-campus shake table has a foot-print of 4.90 m by 3.10 m with a 45-ton payload (Magenes 1989). Specifications of this medium sized shake table include a stroke of ±0.15 m, a peak horizontal velocity of 0.88 m/s, a peak horizontal acceleration of 2.0 g. The frequency bandwidth for this table is 0–15 Hz. For geotechnical testing on this table, a large physical model is used to obtain the necessary data thus a large container is needed to hold the soil and the model. A laminar container was chosen to reduce the energy from the walls, lower boundary effect and to let the soil behave close to its natural condition.

The UCSD medium laminar container (Jakrapiyanun et al. 2001) is composed of 43 frames after recent extension, as the original design was 28 frames. Each frame section is S3 × 5.7 (76.2 mm depth and the weight of the frame 83.2 N/m). The container inside dimensions are 3.90 m long, 1.80 m wide and 2.90 m high after the extension. One of the advantages of the laminar container is that each frame can move relative to the frames above and below it, thus minimizing boundary effect influence. As such, the bearing system of the container consists of 16 cold-rolled steel pipes rolling on stainless steel plates with 1.6 mm thickness. The weight of the steel frame is approximately 10 to 13% of the soil mass depending on the soil density. The total mass of the 43 frames is 4229 kg. The lateral support system consists of 6 columns (W12 × 26). The columns are connected to stiff lower and top frames for rocking protection. This outside frame offers overturning stability and additional stabilization to the moving parts.

The larger off-campus shake table (Englekirk Center) is classified as the largest outdoor shake table in the world. One of the advantages of the outdoor table is that there are no height restrictions, thus models can be as tall as needed. The technical specifications of the shake table (Ozcelik et al. 2008) include a stroke of ±0.75 m, a peak horizontal velocity of 1.8 m/s, a peak horizontal acceleration of 4.2 g for bare table conditions and 1.0 g for a rigid payload of 400 ton, a horizontal force capacity of 6.8 MN, an overturning moment capacity of 50 MN m, and a vertical payload capacity of 20 MN. The frequency bandwidth is 0–20 Hz. Other detailed specifications can be found in Van Den Einde et al. (2004) or the Englekirk website (http://nheri.ucsd.edu).

The large laminar box inside dimensions are 6.75 m length, 3.00 m wide and 4.90 m high. The box is composed of 31 frames, 9 larger bottom frames with W 8 × 31 section and 22 smaller top frames of W 4 × 13 section. Each frame is supported by 18 cold-rolled steel pipes rolling on stainless steel plates with 3.00 mm thickness. This allows the movement of each frame relative to the frames above and below.

This paper presents the recent shake table tests conducted at UCSD and discusses details of the model construction. The performed tests were to evaluate seismic behaviors of various structural models such as pile foundation subjected to liquefaction-induced lateral spreading, retaining walls, tunnels and bucket foundations. In addition to these recent tests, a full-scale wind turbine (Prowell et al. 2009) and other previous tests in the last decade can be found in the UCSD shake table test website (http://nheri.ucsd.edu).

2 Liquefaction-Induced Lateral Spreading Experimentation

Lateral Spreading induced by liquefaction is known to cause excessive movement and possible failure of pile foundation (Yasuda and Berrill 2000). This presents a complex loading situation as the soil undergoes a significant change in its dynamic properties (Finn 2015). The two shake tables due to their ability to simulate ground excitations were used to conduct liquefaction-induced lateral spreading experiments to test their effects on pile foundations. The laminar containers were inclined to allow for the accumulation of lateral deformation.

2.1 Medium Sized Test

Several lateral spreading tests were performed in this experimental program to assess the lateral spreading effect on pile foundation. Testing included different soil profiles, different model inclinations and different pile material. Full details of those tests will be described in Ebeido (2019). One of the tests that were performed on the Powell Lab Shake table is chosen to demonstrate model construction and response. Figure 1 shows the model schematic and instrumentation plan were the response of a reinforced concrete pile is explored. The model was instrumented with various top of the line instrumentation. Sensors were placed in arrays with dense 20 cm spacing to collect comprehensive profile response during shaking. A total number of 160 sensors were installed with data collecting rate of 256 samples per second.

Pore pressure sensors were deployed on both sides of the pile and in the free soil, midway between the pile and container boundary, as well as on both sides of the pile (upslope and downslope). Pressure transducers were installed on both sides of the pile to measure initial static soil pressure during filling and the dynamic pressures during shaking. High sampling accelerometers were used with the ability to record at 25,000 samples per second. With this high sampling rate, the data can be employed to track shear wave velocity changes during seismic events. Strain gauges were installed on the reinforcement bars before concrete casting. Those gauges were densely deployed along the pile reinforcement before concrete casting. Strain data is used to back-calculate bending moment during shaking. Displacement transducers were mounted on the laminar box exterior wall to measure lateral displacements, on the soil surface to measure horizontal and vertical displacements. Pile was also instrumented with transducers to measure pile head displacements above the ground surface.

The box was 3.90 m long, 2.90 m high and 1.80 m wide (inner dimensions) and was inclined at 4° to the horizontal to allow lateral deformation during liquefaction. As noted by (Law and Lam 2001), this configuration essentially simulates a periodic boundary condition. Figure 2a shows a picture of the box placed on the inclined ramp. The profile height of 2.90 m is divided into 3 sand layers with a single pile in the center as shown in Fig. 1. Figure 2b shows the soil surface with the pile in the center after filling. Sand Profile was constructed with a base dense layer, middle loose liquefiable layer and a top dry layer. Sand type used was Ottawa F-65 sand (Bastidas 2016) which is widely used in liquefaction studies.

The base layer, 1.50 m high was constructed by wet sand compaction. Compaction was performed using a plate vibrator in 0.25 m lifts achieving a dense sand density

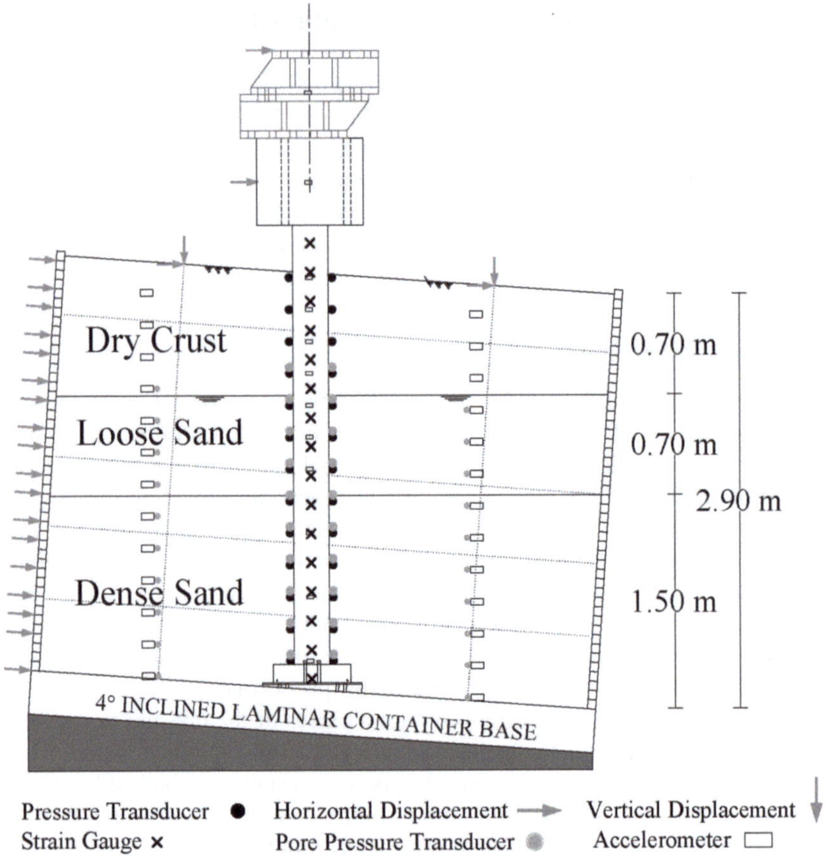

Fig. 1. Schematic Layout of medium size liquefaction-induced lateral spreading experiment

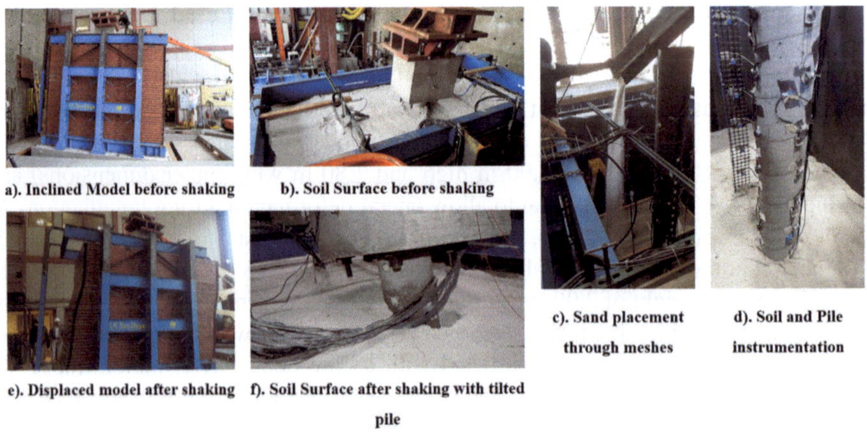

Fig. 2. Medium size liquefaction-induced lateral spreading model

with relative density in the range of 80–90% then saturated after compaction. The middle loose layer was 0.70 m high and an estimated soil relative density in the range of 50–60%. To control the density, meshes (Fig. 2c) were used to allow the sand to separate and settle at a low velocity. The loose layer sand passed through water before settling down to guarantee not having any trapped air. Finally, the top layer of 0.70 m was built dry with some compaction and an estimated soil relative density in the range of 80–90%. An additional mass was placed on the reinforced concrete pile head to add inertial loading to the kinematic lateral spreading load. Figure 2d highlights the instrumentation placed on the pile and the ones placed on the grid to be buried in the soil. The model was subjected to a series of motions, causing the loose layer to liquefy and soil profile to displace. Total box displacement was nearly 1.00 m causing the reinforced pile to break and fall on the soil. Figure 2e and f show the displaced box configuration and gap formed behind the pile when it tilted and rested on the front. 6 cameras were used to record various aspects of the model during shaking events.

2.2 Large Sized Test

Similar to the previous test, a larger model is employed to get a better understanding with a different configuration and a larger scale. The test described in this section is 1 out of 2 tests performed with this size. Full account of those tests will be described in Ebeido (2019). Figure 3 presents the schematic layout of the shake-table experiment, using the large laminar box at the Englekirk Structural Engineering Research Center at the University of California, San Diego. The response of a stiff steel pile subjected to liquefaction-induced lateral spreading is explored with the pile head restraint from movement. The box was 6.75 m long, 4.86 m high and 3.0 m wide (inner dimensions) and was inclined at 4° to the horizontal. As noted by (Law and Lam 2001), this configuration essentially simulates a periodic boundary condition. Approximately 300 sensors were installed in this test, instrumentation was installed along both sides of the pile, along the outside of the box and in the soil between the pile and box boundary as shown in Fig. 3. Sensors installed were the same used in the smaller size test.

The soil profile of 4.86 m was constructed similar to the previous section, in 3 layers using the same Ottawa F-65 sand (Bastidas 2016). A base layer of 1.70 m was compacted in 0.25 m lifts. After compaction, the base layer was saturated then shaken by a series of white noise motions to further densify the sand and achieve a 95–100% relative density. After that, the middle loose sand was constructed by sand deposition through water to give an estimated relative density of 50–60% and a height of 1.50 m. Finally, the top layer of 1.66 m was built dry with some compaction and an estimated soil relative density in the range of 80–90%.

Figure 4a and b show pictures of the box before shaking and the soil surface with a grid painted to easily track surface movement. Figure 4c presents a picture taken by a drone above the box which is another available capability, enabling the photography of hard to reach places of interest. The drone was also used to record the soil surface from above during shaking events. Figure 4d, e and h show various construction stages of instrumented box and pile, compaction of sand layer and placement of sand in the box.

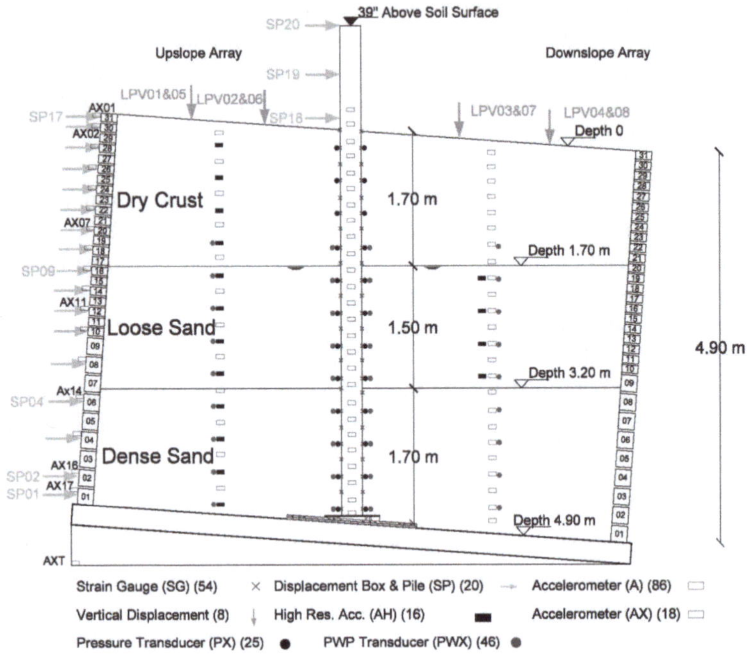

Fig. 3. Schematic Layout of large size liquefaction-induced lateral spreading experiment

Finally, Fig. 4f and g show displaced box configuration after shaking and caving of soil surrounding the pile. In addition to the drone, 12 cameras were used to record various model aspects during shaking events.

Fig. 4. Large size liquefaction-induced lateral spreading configuration

3 Retaining Wall Experimentation

3.1 Bridge Abutment Backwall

Numerous retaining wall tests were conducted on the UCSD shake tables. Figure 5 shows a schematic layout of the set up and instrumentation plan of a bridge abutment backwall in the large laminar container (Wilson and Elgamal 2009, 2010). This configuration was used to evaluate the static and passive pressure on an abutment backwall. The backfill material used was sand from a nearby location. The backwall was installed with a loading system to push the wall passively on the soil. After the wall was placed, the soil was compacted in lifts behind the wall. Construction stages and test setup are shown in Fig. 6. The figure also presents model loading and failure wedge after testing.

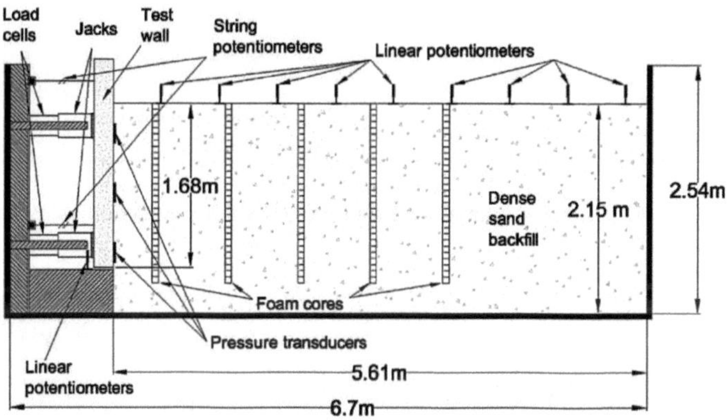

Fig. 5. Schematic Layout of the bridge abutment backwall experiment

4 U-Shaped Retaining Wall Test

The experiment efforts included the use of the large outdoor shake table to evaluate the seismic response of a U-shaped retaining wall. The test configuration was comprised of a U-shaped structure (1 m height, 3 m length, and 1.8 m width) supported on 2.3 m deep compacted backfill and two different backfill conditions adjacent to the structure. Figure 7 shows the construction stages as the laminar box was being built then the supporting soil being compacted. Three test model configurations were considered with different backfill conditions. Two levels of the compaction efforts were considered in terms of a relative density (Dr) measured from a field sand cone test: (1) Dr of 99% as very dense sand; (2) Dr of 85% as dense sand. To minimize any influence of cohesion and water content, clean and dry sand was employed as an additional backfill material. A large number of sensors were installed measuring displacements, strains, accelerations and pressures. Focus was placed on the dynamic earth pressure resulting from earthquake motions and the corresponding wall deflection and bending moments.

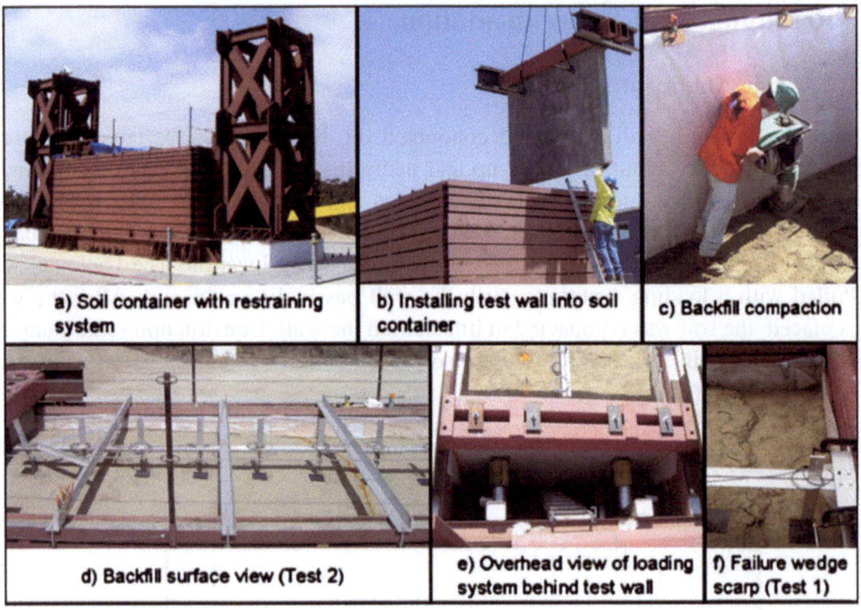

Fig. 6. Bridge abutment backwall model with construction steps

(a) Laminar soil container base (b) Assemblage of laminar soil container (c) Backfill compaction

(d) U-shape structure instrumentation (e) Sand cone test (f) Replacement of backfill

Fig. 7. U-shaped retaining wall model construction

5 Cut and Cover Tunnel Experimentation

A tunnel-soil model was tested using the large-scale outdoor shake table at UCSD to evaluate the seismic response of a cut and cover tunnel. Three test model configurations were employed with different backfill conditions for the surrounding soil and thickness of overburden soil. Two levels of compaction were considered in terms of the relative density of the soil surrounding the tunnel (Dr; about 99% and 85%). The tunnel was placed on 2.9 m deep compacted backfill at about 99% relative density. Soil used in this testing was obtained from a nearby site. Evaluating the seismic response of the tunnel focused on the resulting peak dynamic pressures developed during seismic events and the corresponding deformed structural shape.

Figure 8 shows the tested tunnel model (1 m high and 1.8 m wide). Length of the tunnel (2.8 m) perpendicular to its cross-section was essentially equal to that of the container width (i.e., resulting in a plane strain model configuration). This specimen was designed to represent a 1/9 scale model in terms of the geometric dimensions and the lateral stiffness of the employed steel walls. In prototype scale, the equivalent dimensions for height, width, and length are 9.1 m, 16.2 m, and 25.3 m, respectively. This 1/9 scale tunnel model was also determined on the basis of the 1/3rd scale reinforced concrete (RC) tunnel segment (Kim et al. 2015).

Fig. 8. Cut and cover tunnel configuration (a) tunnel model, (b, c) lowering the model in laminar box, (d) model placed on densified soil

6 Bucket Foundation Experimentation

Aiming to characterize the seismic behavior of suction bucket foundation (also know as: suction caisson), a scaled model of bucket foundation was manufactured and tested at the University of California San Diego. The model was tested at Powell Laboratory 1-g shake table using laminar soil container with saturated dense sand stratum.

Following the scaling laws introduced by Iai and Sugano (1999), the prototype was scaled down by a factor of 1:32. Bucket diameter after scaling is 0.50 m with 0.25 m embedment depth in the sand and 3.40 m shaft height as shown in Fig. 9. The sand stratum was instrumented with pore pressure sensors as the profile was built and compacted in lifts. A detailed description will be discussed in Zayed (2019). Figure 10a shows photos of the model foundation instrumented with pore pressure transducers to measure to the inside and outside pore pressure build-up during the seismic excitation.

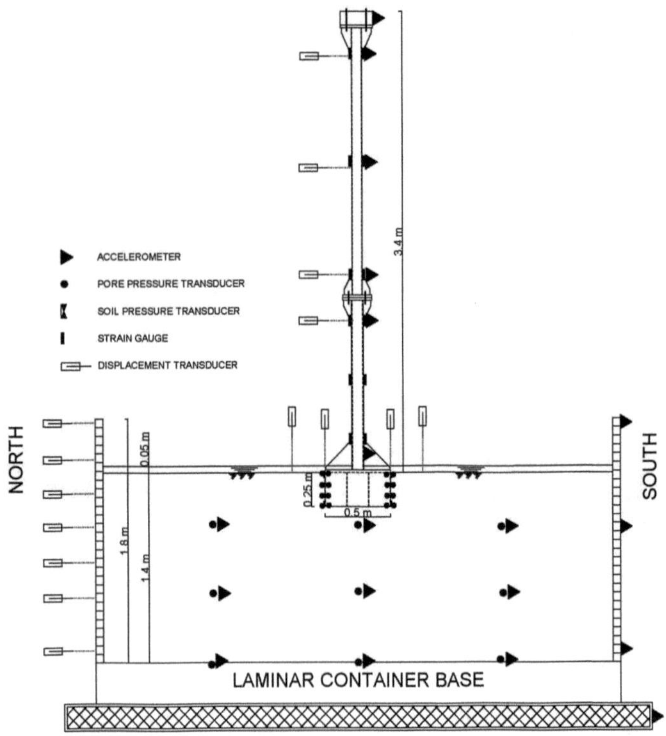

Fig. 9. Schematic Layout for bucket foundation experiment

As shown in Fig. 10a, the inside of the bucket was divided into 6 chambers to simulate the prototype design. A total of 6 vacuum lines were installed to drive the model foundation into the saturated sand deposit, as illustrated in Fig. 10b. Once the vacuum is applied, the pressure inside the bucket, i.e. chambers, is reduced resulting in the bucket being driven into the sand stratum until the top soil surface is in full contact with the bottom face of the bucket top plate. Figure 10c presents the installation process of the bucket foundation into the soil as well as connecting the tower to the shaft. The shake table was used to excite the model with several motions, starting with sinusoidal excitation of 0.1 g and ending with simulated 100% Kobe Earthquake (1995) Takatori record. The response of the bucket foundation was found to adequate

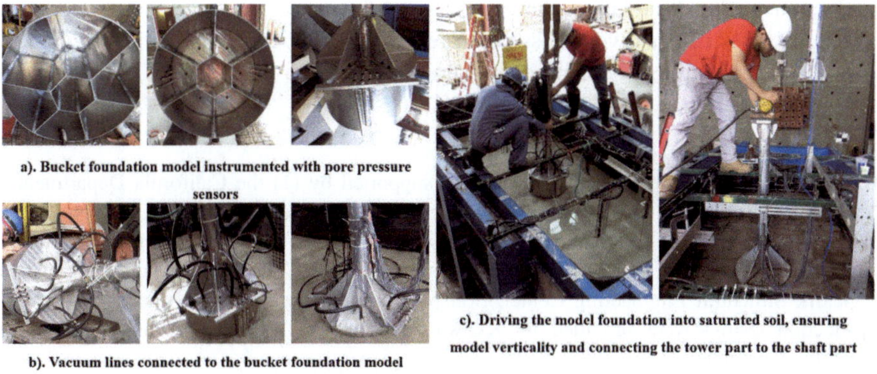

Fig. 10. Bucket foundation model

with medium-strong motions. However, the model experienced large amount of titling when exposed to strong motions.

7 Conclusions

The capabilities present at University of California, San Diego for geotechnical earthquake engineering testing are discussed. The shake table specifications and the laminar containers available are presented with examples of various testing performed at UCSD. The models built are described along with the dense instrumentation used and the construction methodology is discussed.

The large-scale 1-g tests are time consuming and expensive, however they allow the realistic modelling of the problem, densely instrumenting the model and using actual construction material such as reinforced concrete and sand obtained from genuine sites.

Some select findings from the tests described above are:

- For Liquefaction testing during earthquake excitation, dense sand beneath the loose does not liquefy providing support to the pile foundation. High passive pressures develop in the downslope side to resist the pile movement.
- The bridge abutment backwall experiment showed that, passive earth resistance from the dense sand backfill was shown to reach a peak, at a deflection of 2.7 and 3% of the wall height, before decreasing to a residual level in the two tests. The passive failure wedge was observed to be fully formed near the peak measured load.
- Seismic testing on the U-shape retaining wall provided evidence that at high peak accelerations, recorded earth pressures exceed values suggested by Seed and Whitman (1970) by 30%. On the other hand, values recorded were lower than suggestions for low levels of peak acceleration.
- Cut and cover tunnel testing show that overburden pressure can play a role in overall tunnel response. Earth pressure values were lower than design practice values.

- Suction bucket foundation showed good performance against seismic excitation. Large values of negative excess pore pressures can be generated, depending on the level of excitation. These pore pressures are used to stabilize the model against seismic loading.

Acknowledgments. The presented research is supported by (1) the California Department of Transportation with Dr. Charles Sikorsky as the Program Manager, (2) NSF Grant No. OISE-1445712 and (3) Bureau of Reclamation. Their support is gratefully acknowledged. The authors would like to thank the staff and research engineers of the UCSD Powell laboratories and the Englekirk Structural Engineering Center. Soil pressure and Pore Pressure transducers used in liquefaction testing were provided by Kyowa Americas Inc and their technical support is gratefully appreciated.

References

Bastidas, A.M.: Ottawa F-65 Sand Characterization. Ph.D. Dissertation, University of California, Davis (2016)

Ebeido, A.: Lateral-Spreading Effects on Pile Foundations: Large-scale Testing and Analysis. Ph. D. Thesis. Department of Structural Engineering. University of California San Diego, La Jolla, CA (2019)

Englekirk Structural Engineering Center Website: http://nheri.ucsd.edu

Finn, W.L.: 1st Ishihara lecture: an overview of the behavior of pile foundations in liquefiable and non-liquefiable soils during earthquake excitation. Soil Dyn. Earthq. Eng. **68**, 69–77 (2015)

Iai, S., Sugano, T.: Soil-structure interaction studies through shaking table tests. In: Proceedings of the Second International Conference on Earthquake Geotechnical Engineering, Lisboa, Portugal, June (1999)

Jakrapiyanun, W., Ashford, S. Elgamal, A., Meneses, J.: Shake Table Testing using a Laminar Box for Soil-Foundation-Structure Interaction. UCSD Structural Systems Research Project, Report No. SSRP-2001/21 (2001)

Kim, K., Giacalone, C., Elgamal, A., Shing, P.B.: Racking Response of Reinforced Concrete Cut and Cover Tunnel (No. CA15-2420; UCSD/SSRP-15-03). California. Department of Transportation (2016)

Law, H.K., Lam, I.P.: Application of periodic boundary for large pile group. J. Geotech. Geoenvironmental Eng. 889–892 (2001)

Magenes, G.: Design, analysis and calibration of the UCSD shake table. University of California, San Diego, Department of Applied Mechanics/Engineering Sciences (1989)

Ozcelik, O., Luco, J.E., Conte, J.P., Trombetti, T.L., Restrepo, J.I.:. Experimental characterization, modeling and identification of the NEES-UCSD shake table mechanical system. Earthq. Eng. Struct. Dyn. **37**(2), 243–264 (2008). https://doi.org/10.1002/eqe.754

Prowell, I., Veletzos, M., Elgamal, A., Restrepo, J.: Experimental and numerical seismic response of a 65 kW wind turbine. J. Earthq. Eng. **13**(8), 1172–1190 (2009). https://doi.org/10.1080/13632460902898324

Seed, H.B., Whitman, R.V.: Design of earth retaining structures for dynamic loads. In: ASCE Specialty Conference, Lateral Stresses in the Ground and Design of Earth Retaining Structures, pp. 103–147. Cornell Univ., Ithaca, New York (1970)

Ubilla, J., Abdoun, T., Dobry, R.: Centrifuge scaling laws of pile response to lateral spreading. Int. J. Phys. Model. Geotech. **11**(1), 2–22 (2011)

Van Den Einde, L., Restrepo, J., Conte, J.P., Luco, E., Seible, F., Filiatrault, A., Clark, A., Johnson, A., Gram, M., Kusner, D., Thoen, B.: Development of the George E. Brown Jr. network for earthquake engineering simulation (NEES) large high performance outdoor shake table at the University of California, San Diego. In: Proceedings of the 13th World Conference on Earthquake Engineering, Paper No. 3281. Vancouver, BC, Canada (2004)

Wilson, P., Elgamal, A.: Full-scale shake table investigation of bridge abutment lateral earth pressure. Bull. N. Z. Soc. Earthq. Eng. **42**(1), 39 (2009)

Wilson, P., Elgamal, A.:. Large-scale passive earth pressure load-displacement tests and numerical simulation. J. Geotech. Geoenvironmental Eng. **136**(12), 1634–1643 (2010). https://doi.org/10.1061/(ascegt).19-43-5606.0000386

Yasuda, S., Berrill, J.B.: Observations of the earthquake response of foundations in soil profiles containing saturated sands. In: 1st International Conference on Geotechnical and Geological Engineering. Melbourne, Australia, Issue Lecture, pp. 1441–1471 (2000)

Zayed, M.: Large-Scale Seismic Response of Ground and Ground-Structure Interaction Systems. Ph.D. Thesis. Department of Structural Engineering. University of California San Diego, La Jolla, CA (2019)

Response of A 850 KW Wind Turbine Including Soil-Structure Interaction During Seismic Excitation

Muhammad Zayed[1(✉)], Ahmed Elgamal[1], Ghada Saudi[2], Abd el-aziz Khairy Abd el-aal[3], and Khalid El-Zahaby[4]

[1] Department of Structural Engineering, UC San Diego, La Jolla, San Diego, CA 92093-0085, USA
[2] Structures and Steel Construction Department, HBNRC, Giza, Egypt
[3] Head of Egyptian National Seismology Network, NRIAG, Helwan, Egypt
[4] President of Housing and Building National Research Center, Giza, Egypt

Abstract. On-shore wind turbines are typically founded on shallow gravity-based foundations that are designed to transmit vertical and lateral loads. Improved understanding of the foundation, as well as the over-all system behavior will lead to safe and economical designs. In this study, experimental results of a 55 m in-service Gamesa G52/850 wind turbine tower are employed to calibrate a three-dimensional finite element model. The response of a G52/850 wind turbine founded on shallow foundations and subjected to seismic excitation is studied with due consideration of the role of soil-structure interaction. The numerical model accounts for the soil-structure interaction via a pressure dependent multi-yield surface soil constitutive model. The calibrated FE model is then used to investigate the response of the wind turbine under earthquake-like excitation. The fundamental mode shape, the dynamic response as well as the soil-structure interaction effect are reported and evaluated. The investigation provides a valuable insight into the extend that soil-structure interaction influences the behavior of the wind turbine tower under seismic excitation.

1 Introduction

The use of wind energy is increasing rapidly, and the world wind capacity has exceeded the 539 GW milestone in 2017 (REN21 2018). According to the Egyptian Ministry of Electricity and Renewable Energy, the New and Renewable Energy strategy in Egypt aim to increase the share of generated energy from renewable energy sources to 20% out of the total generated energy in Egypt by 2022. Out of the 20% target, 6% is from hydro sources, 12% from wind energy and 2% from other renewable energy sources especially the solar energy. This sets a target of 7.2 GW to be generated from wind energy by 2022 (REN21 2018). Its noticeable that the Egyptian wind market increased from 5 MW in 2001 to 810 MW at the end of 2017 (GWEC 2017). Utility-scale wind turbine plants require a minimum annual average wind speed of 6 m/s. Egypt enjoys excellent wind regimes, particularly in the Suez Gulf where the average

D. Choudhury et al. (Eds.): GeoMEast 2018, SUCI, pp. 114–125, 2019.
https://doi.org/10.1007/978-3-030-01920-4_10

wind speed reaches 10.5 m/s. Investment in the renewable energy increased to nearly USD 2.6 billion in Egypt by end of 2017 (REN21 2018).

The research on seismic performance of wind turbines with shallow foundations is relatively new. Wind turbines have mechanical components such as shafts, links, bearings and gears. These components should operate normally after a design-level earthquake. To ensure the normal operation, wind turbines should satisfy serviceability limit states such as title at the tower top. Generally, wind turbines have longer natural period compared to a building of the same height. The mass at the tower top (nacelle and rotor) could be as large as the mass of the entire tower, which leads to long natural period. Considering the differences in structural characteristics between the wind turbine towers and the traditional buildings, it is necessary to well-understand the dynamic performance of wind turbines when they are subjected to seismic excitations. Although the wind turbine failures due to earthquake are rare (Patil et al. 2016), earthquakes could disturb the normal operation of the whole wind farm for considerable time. Thus, it is of high importance for wind energy developers, insurance companies to ensure the seismic performance of wind turbines constructed in a seismically active regions.

The objective of this paper is to investigate the dynamic performance of a 55 m Gamesa G52/850 wind turbine tower when subjected to seismic excitation. The G52/850 model is a key turbine in the Zafarana wind farm. Zafarana wind farm is located 120 km south of Suez on the Red Sea. Abd el-aal et al. (2016) revealed that the latter zone is considered to be seismically. Special attention is made to the soil-structure interaction of the wind turbine tower. The tower is founded on a shallow square footing that is resting on stiff soil strata. The tower footing as well as the surrounding soil are modeled using three-dimensional finite elements. The model was then subjected to earthquake-like excitation to investigate the seismic response of the wind turbine system including the effects of soil-structure interaction.

2 Literature Review

Veletsos and Meek (1974) illustrated that the dynamic response of a structure on a soft soil due to seismic excitation can be different from the response of a similar structure on a firm soil. Many researches confirmed the latter finding via studies on the effects of soil-structure interaction (SSI) (Adhikari and Bhattacharya 2011; Butt and Ishihara 2012; Harte et al. 2012; He and Li 2008; Hongwang 2012). A full review of the development of SSI was presented by Kausel (2010). Few researchers studied the seismic response of wind turbines without considering the soil structure interaction. Ishihara and Sarwar (2008) assumed that the wind turbine tower is supported on stiff soil with high bearing capacity. Thus, they neglected the soil-foundation-structure interaction. Similarly, Witcher (2005) used the fixity assumption of rigid foundation as the base of the tower. Prowell et al. (2008) conducted a full-scale shake table test to characterize the seismic response of a 23 m tall wind turbine tower.

In his study on the seismic response of tall chimneys, Luco (1986) revealed that the SSI had an effect only for softer soils with shear wave velocity below 750 m/s and it could result in reduction or amplification in the response, depending on the characteristics of the structure, the soil as well as the seismic excitation. Using Monte Carlo simulations, Moghaddasi et al. (2011) observed that the soil-foundation-structure interaction (SFSI), on average, increases the total horizontal displacement of the superstructure with a 10%–30% likelihood of amplification in the structural distortion when accounting for the SFSI for an idealized SDOF structure.

Several efforts have been done to investigate the effect of soil-structure-interaction (SSI) on the seismic structural response of wind turbines. Adhikari and Bhattacharya (2011, 2012) modelled the flexible nature of soil-foundation-structure interaction using rotational lateral springs. This approach was first introduced by Mulliken (1994); Mulliken and Karabalis (1998). Bazeos et al. (2002); He and Li (2008); as well as Hongwang (2012) used discrete springs and dashpots to model the soil at the soil-foundation interface. Harte et al. (2012) adopted different approach where they idealized the SFSI using cone model (Ehlers 1942; Meek and Wolf 1992; Wolf and Deeks 2004). The seismic loading analysis for a 5 MW wind turbine with effects of SSI was studies using finite element model by Prowell et al. (2010). Earlier studies (Bazeos et al. 2002; Lavassas et al. 2003; Prowell et al. 2009) used wind turbine model that lump the nacelle and rotor masses as a point mass at the top of the tower. The available literature related to the dynamic SSI of on-shore wind turbines founded on shallow foundations has been somewhat sparse (Harte et al. 2012).

Change in the soil properties may occur due to the action of cyclic loading resulting from seismic excitation. as a result, the fundamental frequency of the over-all wind turbine system along with the foundation and the supporting soil will change. This change is the system frequency can be critical as it will lead to changes in the wind turbine tower response and in extreme cases mat lead to resonance type of failure. Zhao and Maisser (2006) demonstrated that the fundamental bending frequencies of a wind turbine tower when accounting for the SSI are smaller than that of the case without SSI effects. They found the error to be around 7% between the two cases for the soil conditions considered in their study. Hence, effects of SSI should be given great attention in the design of wind turbines.

3 Finite Element Analyses Framework/Platform

For seismic foundation-ground system analyses, nonlinear constitutive models are needed for the soil and for the pile materials. Applicability to arbitrary cyclic and seismic loading scenarios is also a necessity. In this regard, the open-source computational platform OpenSees (Mazzoni et al. 2006) currently provides such capabilities for the foundation and the soil materials. Nonlinear constitutive models are available in OpenSees (Open System for Earthquake Engineering Simulation) for the steel and concrete (confined and unconfined) materials, as well as for the soil pressure independent and pressure-dependent seismic response (multi-surface plasticity J_2 and Drucker-Prager cone models).

In OpenSees, dry and saturated coupled solid–fluid two-dimensional (2D) plane strain (Yang et al. 2004) and 3D brick (Yang 2000; Yang and Elgamal 2002) soil elements are available following the Chan (1988) original u–p formulation. In addition, incremental-plasticity multi-yield surface (Mroz 1967; Prevost 1985) soil models have been developed and employed for conducting seismic analyses (Elgamal et al. 2009;

Yang et al. 2003). In the pressure-dependent models (Elgamal et al. 2003; Yang and Elgamal 2002), emphasis was placed on simulating the liquefaction-induced shear strain accumulation mechanism in clean medium-dense sands. As such, the aforementioned soil elements and models allow for simulation of dry and fully saturated soil conditions.

The availability of these advanced algorithms in OpenSees allows for general purpose 3D seismic modeling environments. In order to facilitate the pre- and post-processing efforts for 3D pile-ground analyses, a Windows-based graphical interface OpenSeesPL (Lu et al. 2006) is being developed to simplify the FE analyses for pile systems, that being an important 3D practical application in geotechnical engineering. OpenSeesPL (http://soilquake.net/openseespl/) allows for: (i) convenient generation of the mesh and associated boundary conditions and loading parameters (FE input file), (ii) execution of the computations using the OpenSees platform, and (iii) graphical display of the results for the pile and the ground systems.

In the following sections, a three-dimensional finite element model is developed for an in-service wind turbine tower along with its shallow foundation and the surrounding soil domain. The model is then calibrated to be in a good agreement with the experimental field measurements reported by Saudi (2017), with main focus on matching the first fundamental frequency measured in the field.

3.1 Model Description

The wind turbine tower presented in this study is modeled after the Gamesa G52/850 KW model (Fig. 1a). The tower model was simplified by using the equivalent mass for the nacelle and rotor assembly and lumping it at the top of the tower. Such assumption was revealed to be quite accurate when lower modes govern the seismic response (Prowell et al. 2009). The turbine tower consists of three hollow tubular steel sections bolted together with connection flanges. Each section is made of several sub-sections welded together. The hub height of the turbine is 55 m with a rotor diameter of 52 m. Thickness of the tower varies across the height. The maximum thickness is found to be 26 mm and the minimum thickness is 12 mm. The total top mass including the hub, rotor and nacelle is 43 tons, and the tower mass is 62 tons. The tower was divided into 11 beam-column elements with a flexural stiffness based on the cross section of the tower at the center of each element (Jonkman et al. 2009; Prowell et al. 2010).

In this study, an actual turbine foundation is modeled. The shallow foundation is a reinforced concrete block that is idealized to be 8.3 m $\times$ 8.3 m with height of 1.2 m. The turbine tower is connected to the foundation via means of un-bonded post tensioned bars that extend to the bottom of the foundation, and it is modeled as rigid connection.

3.2 Calibration of Fixed-Base Finite Element Model

A simple fixed-base finite element model was developed that represents the turbine tower, nacelle and rotor through beam-column elements. Earlier studies revealed that the use of beam-column elements to model the turbine tower results in quite accurate results compared with more detailed shell elements models (Bazeos et al. 2002).

(a)

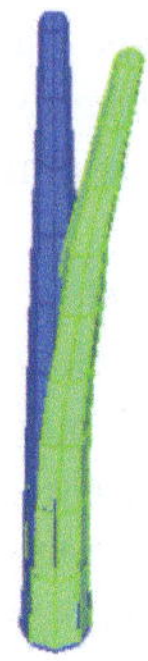

Fundamental mode shape as identified experimentally via modal identification
f = 0.552 Hz

(b)

Fig. 1. a) Gamesa G52/850 wind turbine tower at Zafarana wind farm, and b) Identified structural mode shape of G52/850 wind turbine tower from model identification "After Saudi (2017)"

The latter simplified configuration represents the most commonly used approach for modeling seismic response of wind turbine (Bazeos et al. 2002; Jonkman and Buhl Jr. 2005; Witcher 2005).

With a Young's Modulus for steel of 210 GPa, the fixed base model closely matches the relevant first fundamental frequency reported by Saudi (2017). The first bending fundamental frequency for an in-service wind turbine tower of the same model is measured via model identification and found to be 0.552 Hz (Fig. 1b) while the obtained frequency from the FE model is 0.547 Hz, with error 0.98%.

3.3 Soil-Structure Interaction Finite Element Model

In this section, a three-dimensional finite element mesh is used to represent the soil domain surrounding the tower foundation. This approach is more realistic to model the foundation rather than using equivalent springs and dashpots, especially if the soil profile is layered. The numerically consider the effect of SSI for the modeled wind turbine, the fixed base model was extended using OpenSeesPL (Lu et al. 2006) to include the soil domain with the foundation model. Non-linear 8-node brick elements with appropriate stiffness and density are used to model the foundation and the soil. The soil model consists of three layers; wadi deposit, silty clay and clay stone. The soil profile is described in details by El-Zahaby and Elgamal (2015). Pressure-Dependent-Multi-Yield Surface constitutive model is employed to represent the soil profile. Full description of the material properties used with the numerical model is reported by Zayed (2019).

The reference shear wave velocity for the three soil strata are 420, 510 and 590 m/s, respectively. Aiming to minimize the boundary effects, the soil domain extended to depth of 20 m and horizontal distance 20 m after the foundation. The model consisted of over 7,000 soil elements. Figure 2a shows the FE soil mesh with the shallow foundation and the idealized turbine tower. More refined meshes were investigated to ensure that the reported results are not influenced by the soil model geometry and discretization.

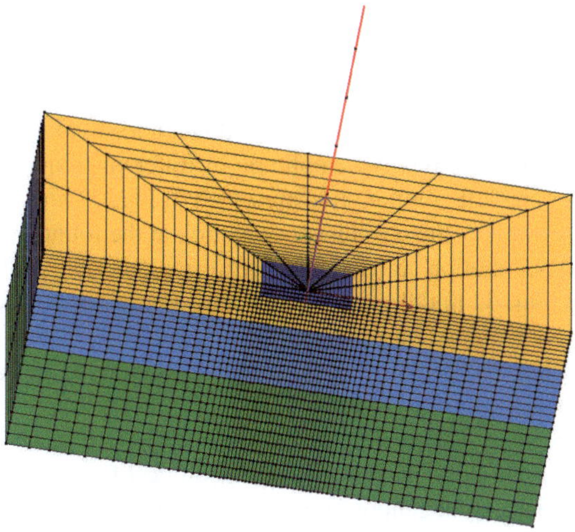

(a)

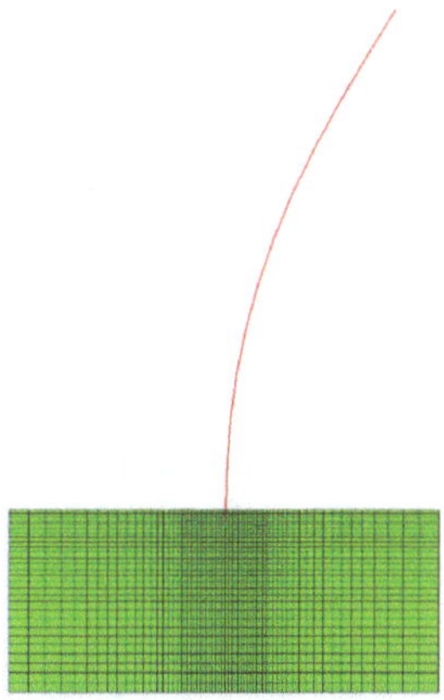

(b)

Fig. 2. a) Soil mesh, shallow foundation and idealized wind turbine tower, and b) Fundamental bending mode shape of the wind turbine tower using the actual soil profile

Using soil mesh only without any structure, the first resonant frequency of the soil model was evaluated numerically from the FE model and found to be 6.4 Hz. An approximation of the studied soil model fundamental frequency is found to be 6.25 Hz using the equation provided by Kramer (1996), assuming half space, uniform and homogenous soil layer. Attention was given to make sure that the resulting resonance frequency of the soil mesh does not coincide with the fundamental frequency of the wind turbine tower. However, it can still excite higher modes, depending on the seismic excitation.

When modeled with the soil mesh, the first fundamental frequency occurred at 0.543 Hz. When compared with the experimental measured frequency, the resulting error is 0.43%, which is half the error of the fixed base case. Figure 2b shows the corresponding mode shape for the fundamental frequency resulting from modeling the SSI. The resulting fundamental frequency from the fixed base case and the SSI case are quite close, i.e. 0.547 Hz vs. 0.473 Hz, respectively. This can be explained by the nature of the soil profile at this specific site, which is considered to be firm ground.

3.4 Earthquake Loads

A record of the 1994 Northridge Earthquake was used as the input motion at the base of the SSI model. The SSI model was then subjected to the 1994 Northridge record (6.7 moment magnitude) as recorded at the Rinaldi receiving station. Since the earthquake record was recorded at the ground surface, the motion was decided to be used as the excitation input at the model base with amplitude factor 0.5.

4 Simulation Results and Discussion

With a relatively low damping ratio of 2% employed for the soil domain in simulations, the soil layer both amplified and altered the relative amplitude of different frequency components from the input motion. The left column plots presented in Fig. 3 show the input excitation at the model base as well as the absolute acceleration at the foundation level. Frequency spectra of the input motion as well the propagated motion at the foundation level are presented in Fig. 3, middle column plots. Comparing the two spectra, a transfer function for the soil model can be obtained. The transfer function presented in the most right plot of Fig. 3 illustrates which frequencies were amplified and which frequencies were decayed. It can be noted that peak of the transfer function occurs at frequency 6.62 Hz, which is the fundamental frequency of the soil strata. The latter fundamental frequency shows a good agreement with the theoretical fundamental frequency of a half-space, uniform and homogenous soil layer (Kramer 1996) that is found to be 6.25 Hz. It also shows a good agreement with the fundamental frequency obtained from modeling the soil strata only without any structure.

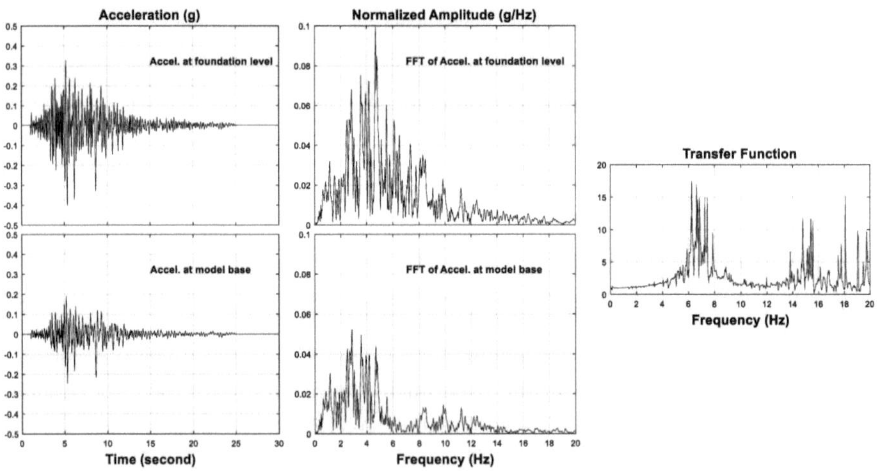

Fig. 3. Time history plots and FFT of the input excitation at model base as well as the propagated motion at the foundation level, and the transfer function between the two locations

To investigate the implications of the seismic excitation on the tower response, time history plots for the top tower acceleration and displacement as well as rotation are presented in Fig. 4. The acceleration at the top tower reached a maximum value of 0.39 g, at time 5.4 s. From the middle plot in Fig. 4, it can be observed that the tower experienced maximum lateral displacement of 0.07 m at 9.2 s, just after the strong part of the base excitation motion previously presented in Fig. 3. At 9.2 s, the maximum rotation of the top tower was observed and found to be 0.0033 rad, i.e. 0.19°.

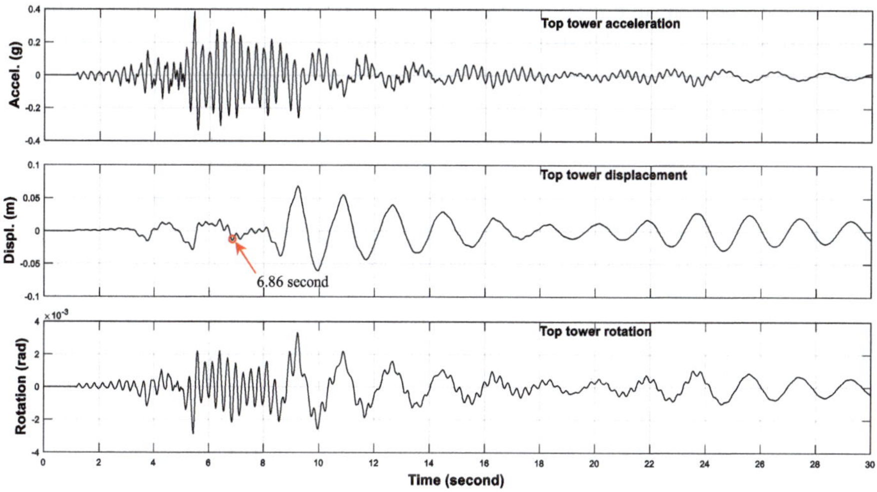

Fig. 4. Time history plots of top tower acceleration, displacement and rotation

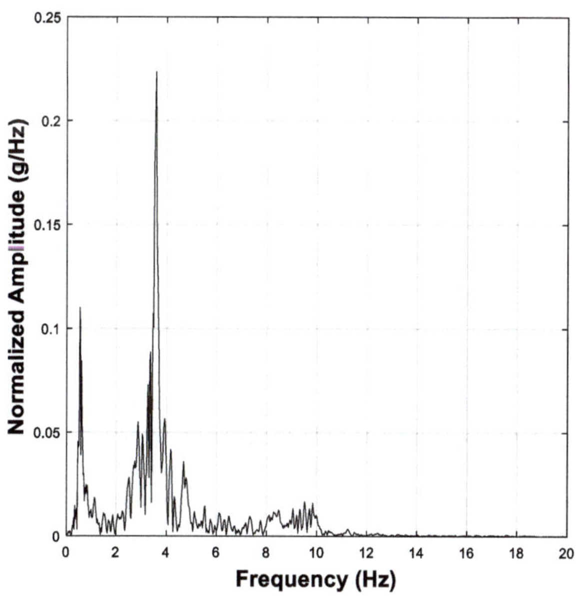

Fig. 5. Frequency spectrum of the top tower acceleration

The top tower response presented in Fig. 4 shows high frequency component at early time of shaking. The higher frequency component can be related to higher mode shapes of the wind turbine tower. Figure 5 shows frequency spectrum of the top tower acceleration, where the contribution of the high frequency component, i.e. 3.56 Hz, is larger than the contribution of the fundamental frequency, i.e. 0.54 Hz. This higher frequency of 3.56 Hz shown in the top tower response is most likely representing the second mode shape. The latter conclusion can be confirmed by plotting the lateral deformation of the tower at early time of shaking. For instance, the lateral deformed shape of the tower at time 6.86 s presented in Fig. 6 shows good agreement with the commonly known 2nd model shape of a SDOF structure.

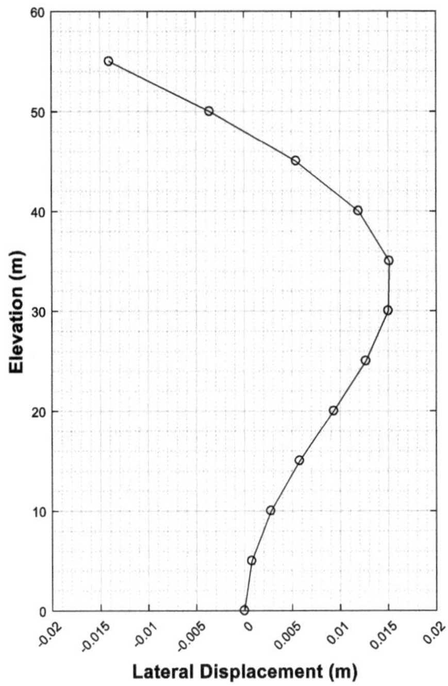

Fig. 6. Lateral deformed shape of the wind turbine tower at time 6.84 s (see top tower displacement time history presented in Fig. 4)

5 Conclusions

Numerical investigation of the resonant characteristics of 850 KW wind turbine tower at Zafarana wind farm is demonstrated. Finite element modeling showed that the relatively stiff soil profile at the wind farm site resulted in a small SSI effect on the fundamental mode. The ground motion of the1994 Northridge earthquake record was found to have energy that helped to excite the higher modes of the wind turbine tower under investigation. Hence, it can be concluded that there is a likelihood that higher modes can play significant and important role, especially with the stiff soil profile presented in this study. However, such effect in the dynamic response of the turbine tower is small in comparison to the safety margins used to isolate the resonant frequencies of the turbine during operation and may not require redesign of the turbine to account for SSI effects.

Acknowledgement. The authors would like to express their gratitude to all the organizations, corporations, and individuals who contributed to this investigation. The research is funded by the US-Egypt Cooperative Research Project, entitled: "*Seismic Risk Assessment of Wind Turbine Towers in Zafarana Wind Farm, Egypt*" (NSF grant No. OISE 1445712).

References

Abd el-aal, A.K., Yagi, Y., Kamal, H.: Implementation of integrated multi-channel analysis of surface waves and waveform inversion techniques for seismic hazard estimation. Arab. J. Geosci. **9**(4), 322 (2016). https://doi.org/10.1007/s12517-016-2329-6

Adhikari, S., Bhattacharya, S.: Vibrations of wind-turbines considering soil-structure interaction. Wind Struct. **14**(2), 85 (2011)

Adhikari, S., Bhattacharya, S.: Dynamic analysis of wind turbine towers on flexible foundations. Shock Vibr. **19**(1), 37–56 (2012)

Bazeos, N., Hatzigeorgiou, G., Hondros, I., Karamaneas, H., Karabalis, D., Beskos, D.: Static, seismic and stability analyses of a prototype wind turbine steel tower. Eng. Struct. **24**(8), 1015–1025 (2002)

Butt, U.A., Ishihara, T.: Seismic load evaluation of wind turbine support structures considering low structural damping and soil structure interaction. Eur. Wind Energy Assoc. Ann. Event **16–19**, 04 (2012)

Chan, A.H.-C.: A unified finite element solution to static and dynamic problems of geomechanics (1988)

Ehlers, G.: The effect of soil flexibility on vibrating systems. Beton und Eisen **41**(21/22), 197–203 (1942)

El-Zahaby, K., Elgamal, A.: Seismic Risk Assessment of Wind Turbine Towers in Zafarana Wind Farm Egypt (Progress Report No. 2). Retrieved from Housing & Building National Research Center (HBRC), Giza, Egypt (2015)

Elgamal, A., Lu, J., Yang, Z., Shantz, T.: Scenario-focused three-dimensional computational modeling in geomechanics. Paper presented at the Proceedings of the 4th International Young Geotechnical Engineers Conference (2009)

Elgamal, A., Yang, Z., Parra, E., Ragheb, A.: Modeling of cyclic mobility in saturated cohesionless soils. Int. J. Plast **19**(6), 883–905 (2003)

GWEC: Global Wind Report: Annual Market Update 2017 (2017). http://files.gwec.net/files/GWR2017.pdf

Harte, M., Basu, B., Nielsen, S.R.: Dynamic analysis of wind turbines including soil-structure interaction. Eng. Struct. **45**, 509–518 (2012)

He, G., Li, J.: Seismic analysis of wind turbine system including soil-structure interaction. Paper presented at the Proceedings of the 14th World Conference on Earthquake Engineering (2008)

Hongwang, M.: Seismic analysis for wind turbines including soil-structure interaction combining vertical and horizontal earthquake. Paper presented at the 15th World Conference on Earthquake Engineering, Lisbon, Portugal (2012)

Ishihara, T., Sarwar, M.: Numerical and theoretical study on seismic response of wind turbines. Paper presented at the European wind energy conference and exhibition (2008)

Jonkman, J., Butterfield, S., Musial, W., Scott, G.: Definition of a 5-MW reference wind turbine for offshore system development (2009)

Jonkman, J.M., Buhl Jr, M.L.: Fast user's guide-updated, August 2005

Kausel, E.: Early history of soil–structure interaction. Soil Dyn. Earthq. Eng. **30**(9), 822–832 (2010)

Kramer, S.L.: Geotechnical earthquake engineering. In: Prentice–Hall International Series in Civil Engineering and Engineering Mechanics. Prentice-Hall, New Jersey (1996)

Lavassas, I., Nikolaidis, G., Zervas, P., Efthimiou, E., Doudoumis, I., Baniotopoulos, C.: Analysis and design of the prototype of a steel 1-MW wind turbine tower. Eng. Struct. **25**(8), 1097–1106 (2003)

Lu, J., Yang, Z., Elgamal, A.: OpenSeesPL three-dimensional lateral pile-ground interaction version 1.00 user's manual. Rep. No. SSRP-06, 3 (2006)

Luco, J.E.: Soil-structure interaction effects on the seismic response of tall chimneys. Soil Dyn. Earthq. Eng. **5**(3), 170–177 (1986)

Mazzoni, S., McKenna, F., Scott, M.H., Fenves, G.L.: The open system for earthquake engineering simulation (OpenSEES) user command-language manual (2006)

Meek, J.W., Wolf, J.P.: Cone models for homogeneous soil. I. J. Geotech. Eng. **118**(5), 667–685 (1992)

Moghaddasi, M., Cubrinovski, M., Chase, J.G., Pampanin, S., Carr, A.: Effects of soil–foundation–structure interaction on seismic structural response via robust Monte Carlo simulation. Eng. Struct. **33**(4), 1338–1347 (2011). https://doi.org/10.1016/j.engstruct.2011.01.011

Mroz, Z.: On the description of anisotropic workhardening. J. Mech. Phys. Solids **15**(3), 163–175 (1967)

Mulliken, J.S.: Discrete models for foundation-soil-foundation interaction in time domain. University of South Carolina (1994)

Mulliken, J.S., Karabalis, D.L.: Discrete model for dynamic through-the-soil coupling of 3-D foundations and structures. Earthq. Eng. Struct. Dyn. **27**(7), 687–710 (1998)

Patil, A., Jung, S., Kwon, O.-S.: Structural performance of a parked wind turbine tower subjected to strong ground motions. Eng. Struct. **120**, 92–102 (2016)

Prevost, J.H.: A simple plasticity theory for frictional cohesionless soils. Int. J. Soil Dyn. Earthq. Eng. **4**(1), 9–17 (1985)

Prowell, I., Elgamal, A., Lu, J.: Modeling the influence of soil structure interaction on the seismic response of a 5 MW wind turbine (2010)

Prowell, I., Veletzos, M., Elgamal, A., Restrepo, J.: Shake table test of a 65 kW wind turbine and computational simulation. Paper presented at the 14th World Conference on Earthquake Engineering, Beijing, China (2008)

Prowell, I., Veletzos, M., Elgamal, A., Restrepo, J.: Experimental and numerical seismic response of a 65 kW wind turbine. J. Earthq. Eng. **13**(8), 1172–1190 (2009)

REN21: Renewables 2018 Global Status Report. Retrieved from REN21 Secretariat, Paris, France (2018). http://www.ren21.net/wp-content/uploads/2018/06/17-8652_GSR2018_FullReport_web_-1.pdf

Saudi, G.: Experimental modal identification of full-scale wind turbine towers. Paper presented at the 7th International Operational Modal Analysis Conference, Ingolstadt, Germany, 10–12 May 2017

Veletsos, A.S., Meek, J.W.: Dynamic behaviour of building-foundation systems. Earthq. Eng. Struct. Dyn. **3**(2), 121–138 (1974)

Witcher, D.: Seismic analysis of wind turbines in the time domain. Wind Energy **8**(1), 81–91 (2005)

Wolf, J.P., Deeks, A.J.: Foundation Vibration Analysis: A Strength of Materials Approach. Elsevier (2004)

Yang, Z.: Numerical modeling of earthquake site response including dilation and liquefaction (2000)

Yang, Z., Elgamal, A.: Influence of permeability on liquefaction-induced shear deformation. J. Eng. Mech. **128**(7), 720–729 (2002)

Yang, Z., Elgamal, A., Parra, E.: Computational model for cyclic mobility and associated shear deformation. J. Geotech. Geoenviron. Eng. **129**(12), 1119–1127 (2003)

Yang, Z., Lu, J., Elgamal, A.: A web-based platform for computer simulation of seismic ground response. Adv. Eng. Softw. **35**(5), 249–259 (2004)

Zayed, M.: Large-Scale Seismic Response of Ground and Ground-Structure Interaction Systems. Ph.D. Thesis. Department of Structural Engineering. University of California San Diego, La Jolla, CA (2019)

Zhao, X., Maisser, P.: Seismic response analysis of wind turbine towers including soil-structure interaction. Proc. Inst. Mech. Eng. [K]: J. Multi-body Dyn. **220**(1), 53–61 (2006)

Effect of Gradation on the Dynamic Response of Sands

Devi Priyanka Darisi[1(✉)], Beeram Ajay Reddy[2], and Sireesh Saride[1]

[1] Department of Civil Engineering, IIT Hyderabad, Hyderabad, India
{ce16resch11002, sireesh}@iith.ac.in
[2] Irrigation and CAD Department, Government of Telangana, Hyderabad, India
beeramajayreddy@gmail.com

Abstract. This paper presents the dynamic soil parameters of sand viz., shear modulus, and damping ratio, which are needed to be considered for the structural design in the areas susceptible to cyclic loading. The aim of the research is to relate the gradation of sands with dynamic properties at small strains. These properties, in fact, are the function of soil characteristics, loading conditions and the mean effective principal stress at any instant. It was witnessed from the stiffness degradation and damping characteristics of sand that the gradation significantly influences the stiffness due to alteration in the stress state of the soil. Furthermore, the data also revealed that the threshold shear strain at the onset of modulus degradation slightly reduces with increase in coefficient of uniformity.

Keywords: Dynamic soil properties · Coefficient of uniformity
Ennore sand · Resonant column testing

1 Introduction and Background

Since 1970s, it has been globally recognized that the stress-strain response of soils and soft rocks is highly non-linear. This element has led to the evolution of various design methods that account for stiffness non-linearity. Nowadays, the described phenomenon is routinely incorporated into numerous computer codes. Besides these, many in-situ and laboratory testing processes have been developed, which supports in the examination of the stress-strain response, even at small strains (Viana da Fonseca et al. 2009).

This non-linear soil response of the shear stiffness is due to dependency on the shear strain (γ) and other parameters e.g. specimen density, effective confining stress etc. However, the response at strains lower than 10^{-3}% can be considered practically elastic and hence, the resulting modulus is treated as the maximum shear modulus (G_{max}). This knowledge about G_{max} is important principally in two areas i.e., seismic ground response analysis and liquefaction assessment (Andrus and Stokoe 2000; Yang and Yan 2009). It also plays a vital role in excavations and foundations, pavements, offshore marine structures etc. (Atkinson 2000). Currently, direct calculation of seismic wave velocities in the laboratory has become a common global practice. It gives a powerful potential in the small strain stiffness determination, which was later recognized as a reference soil characteristic. Among various laboratory techniques available for measurement of shear stiffness, bender element and resonant column (RC) tests are

D. Choudhury et al. (Eds.): GeoMEast 2018, SUCI, pp. 126–139, 2019.
https://doi.org/10.1007/978-3-030-01920-4_11

recognized to be more reliable. Bender elements were actually developed for soft soils. They enable transmitting and receiving of either shear/compression wave and can be installed virtually in any conventional soil testing apparatus. More facts about this technique are presented in Ferreira et al. 2011, whereas details of RC procedure are described in the latter part of this paper. Both methods have different benefits and limitations. For example, the elucidation of RC data has been well established, whereas the reading of bender element signals is one way difficult and involves uncertainty (Lee and Santamarina 2005).

1.1 The Significance of Gradation in Dynamic Soil Studies

In spite of the above-mentioned parameters influencing the dynamic behavior of soil, it was interestingly found that the soil gradation also plays a key role. It is a basic soil parameter which primarily depends on particle size and their proportion. The reason for accounting the influence of gradation was to address the spatial and temporal variability in a given site under given loading condition. The spatial (with distance and depth) and temporal variability in the sizes of the particles which may affect the inter-particle stress distribution leading to alteration of dynamic response. The temporal variability is a minor factor, as only in few cases loading results in breakage of particles eventually with time leading to change in the soil gradation. This role of loading is seen in case of heavy structures lying on weak foundations or weak rocks. Therefore, to account for this variability in gradation with distance and depth at a site, it is necessary to consider the effect of gradation on dynamic soil properties.

The studies of Silver and Seed (1971), Hardin and Drnevich (1972b), Dobry and Vucetic (1987), Ravishankar et al. (2005) etc. concluded that with increase in relative density and confining pressure, shear modulus increases and damping decreases for soils such as sand, clay, and gravel at a particular shear strain. Whereas, with an increase in cyclic strain, shear modulus decreases and damping ratio increases due to increased mobility of particles. However, very few researchers have considered the effect of soil gradation on dynamic soil properties. Hence, there is a research need to understand the dynamic behaviour of sands with different gradation.

In order to account for the gradation effect, the soil parameters such as uniformity coefficient (C_u), curvature coefficient (C_c), effective size, mean particle size etc. that vary with the soil gradation shall be considered. As the C_u and C_c are the key parameters which primarily decides whether the soil is well graded or poorly graded, in the present study uniformity coefficient ($C_u = d_{60}/d_{10}$) was considered to represent the effect of gradation.

There are certain empirical equations given by researchers for stiffness degradation (G/G_{max}) and damping ratio based on tests on granular soils with uniform gradation curves. Nevertheless, it was not experimentally verified that those equations are applicable to well-graded cohesionless soils. Wichtmann and Triantafyllidis (2013) have proven that the existing empirical equations for G_{max} (such as formulae proposed by Hardin and Black (1966), Iwasaki and Taksuoka (1977) and so on), do not describe the effect of uniformity coefficient on dynamic properties. It was also seen that for a constant void ratio and mean confining pressure, G_{max} decreases significantly with uniformity coefficient. To account for this influence, a correlation was proposed by

Wichtmann and Triantafyllidis (2013) for G_{max} as a function of C_u. In view of the above mentioned variability in the behaviour of sands, an attempt was made to investigate the influence of C_u on dynamic behaviour of graded sands.

2 Material Characterization

Graded Ennore sand used for the present study was wet-sieved to remove existing fines, if any. The basic properties of the sand such as specific gravity (G_s), minimum void ratio (e_{min}), maximum void ratio (e_{max}), were then determined as per IS 2720-part III, IV and XIV code. The gradation of soil samples was performed using Ennore sand of different grades to obtain a C_u in the range of 1 to 6. In case of lower C_u, only one grade sand was used. Whereas, for higher C_u values, different grade sands were mixed in different proportions. The gradation curves and the basic properties of sand are as shown in Fig. 1 and Table 1, respectively.

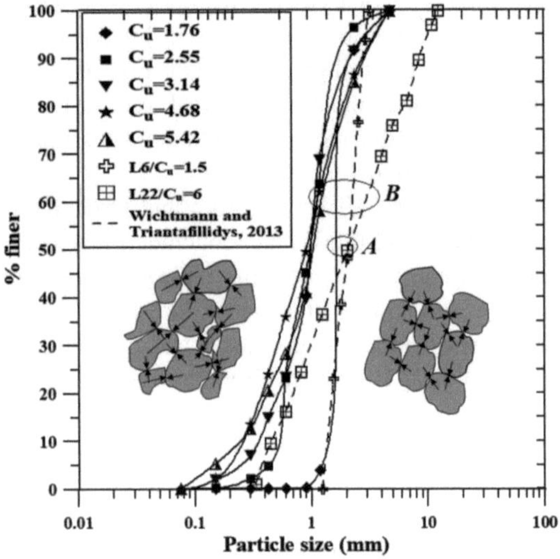

Fig. 1. Grain size distribution curves of Ennore sand with different C_u values

3 Specimen Preparation

A cylindrical specimen of size 50 mm diameter and 100 mm length was prepared in a split mould apparatus. The quantity of sand required for each test was calculated depending on the desired relative density. Using a 150 g tamping rod, sand was compacted in 5 layers into a split mould with rubber membrane after removing the entrapped air. The number of blows required for each layer was determined using few trial samples based on chosen relative density. The top cap was fitted after careful leveling of topsoil surface. By opening the drain valve a small back pressure of 20 kPa

Table 1. Basic properties of sand with different uniformity coefficient values

C_u	G_s	e_{min}	e_{max}
1.76	2.61	0.67	0.891
2.34	2.632	0.662	0.874
3.14	2.635	0.591	0.829
4.68	2.642	0.534	0.768
5.42	2.67	0.497	0.727
6.54	2.68	0.486	0.709
7.6	2.66	0.493	0.711
8.41	2.70	0.493	0.714

(negative) was applied, to keep the specimen intact. The suction was then stopped and the split mould was gently dismantled.

4 Testing

To determine the shear modulus (G) and damping properties of sand under different dynamic conditions, fixed-free RC tests were carried as per ASTM D4015-92 (reapproved 2000) code (Fig. 2). The shear modulus and damping ratio (D) of the specimen were calculated by means of Eqs. 1, 2 and 3 with the help of resonant frequency and free-vibration decay curve.

Fig. 2. Steps involved in sand sample preparation for RC test

$$G = \rho(v_s)^2 \tag{1}$$

$$v_s = 2\pi f_r l/\beta \tag{2}$$

$$D = 1/2\pi n\left(\ln\frac{z_0}{z_n}\right) \tag{3}$$

where, ρ is the initial density of the specimen; v_s is the shear wave velocity; f_r is the fundamental frequency of soil specimen; β is the factor dependent on the inertia of specimen and drive system; l specifies length of the sample; n is number of cycles; z_0 and z_n are vibration amplitudes after excitation is shut off.

5 Results and Discussion

The present study investigates the influence of uniformity coefficient on dynamic properties through a series of RC tests on sand samples with varied C_u values in the range of 1 to 6. The results from the tests performed at different confining pressures, shear strains under constant relative density of 50% are discussed below.

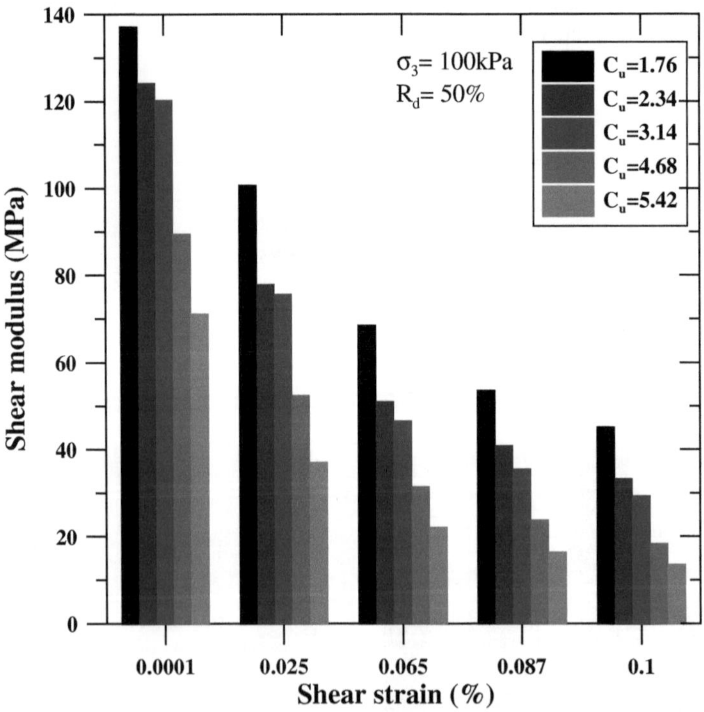

Fig. 3. Comparison of shear modulus vs. shear strain for different coefficient of uniformity at 100 kPa confining pressure and 50% relative density

5.1 Shear Modulus

Tests were performed at 50% relative density of sand for varied coefficients of uniformity (C_u = 1 to 6) and shear strains in the range of 0.0001% to 0.1%. The variation of the secant shear modulus of sand at shear strains of 0.0001%, 0.025%, 0.065%, 0.087%, 0.1% for confining pressures of 100 kPa and 600 kPa are as shown in Figs. 3 and 4, respectively. From Fig. 3, at a stress of 100 kPa and C_u of 1.76, shear modulus of 137 MPa, 100.6 MPa, 68.4 MPa, 53.4 MPa and 45 MPa were observed for corresponding shear strains of 0.0001%, 0.025%, 0.065%, 0.087%, and 0.1%, respectively. Similar trend was noticed at a higher confining pressure of 600 kPa. From the Fig. 4, for same C_u and shear strains, shear modulus of 312 MPa, 283.1 MPa, 227.6 MPa, 202.5 MPa, 189.5 MPa were witnessed. This shows an increase in the modulus at higher confining stresses. This is due to an increased shear stiffness of the specimen.

Also, there is a reduction in the modulus as shear strain moves from 0.0001% to 0.1%. This is due to increased mobility of the soil particles. The data also shows that for a confining stress of 100 kPa and shear strain of 0.0001%, shear modulus of 137 MPa, 124 MPa, 120.1 MPa, 89.4 MPa, 71 MPa were observed for C_u values of 1.76, 2.55, 3.14, 4.68 and 5.42, respectively. A similar effect of uniformity coefficient was noticed at higher confining stress of 600 kPa. Usually, it is believed that well graded soils possess more strength compared to uniformly graded (or poorly graded) soils as maximum void space in the soils is filled by small size particles. However, it

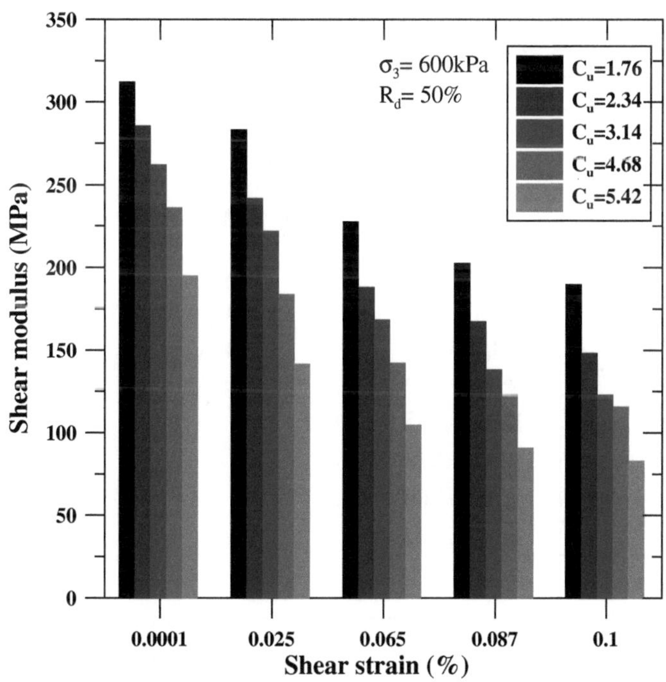

Fig. 4. Comparison of shear modulus vs. shear strain for different coefficients of uniformity at 600 kPa confining pressure and 50% relative density

was explicitly noticed that irrespective of the shear strain amplitude and confining pressure, there was a reduction in shear modulus with a rise in uniformity coefficient. It is because, as C_u increases, the gradation of the soil changes. As the number of particles with different sizes increases the contact area between the particles increases, resulting in an uneven stress distribution, as explained pictorially in Fig. 1. Hence, stiffness reduces with uniformity coefficient.

However, each soil possesses different shear modulus under different conditions of confining pressure, initial void ratio, shear strain, plasticity index, over consolidation ratio etc. Hence, Zen et al. (1978), Dobry and Vucetic (1987), Ishibashi and Zhang (1993) and several others have carried out tests on various soils and generalized the modulus variation with the aid of a normalized parameter, G/G_{max}. They also concluded that the shape of average modulus degradation curve (G/G_{max} vs. γ) remains same except the magnitude of G/G_{max} at various strains. Therefore, the variation of G/G_{max} with shear strain was studied considering the effect of C_u, as demonstrated in latter part.

5.2 Stiffness Degradation and Threshold Shear Strain

It was noticed from the data that G/G_{max} vs. shear strain variation is similar to the variation of shear modulus vs. γ, as explained in Figs. 5 and 6. At a cyclic shear strain of 0.025% and confining stress of 100 kPa, with an increase in C_u from 1.76 to 5.42,

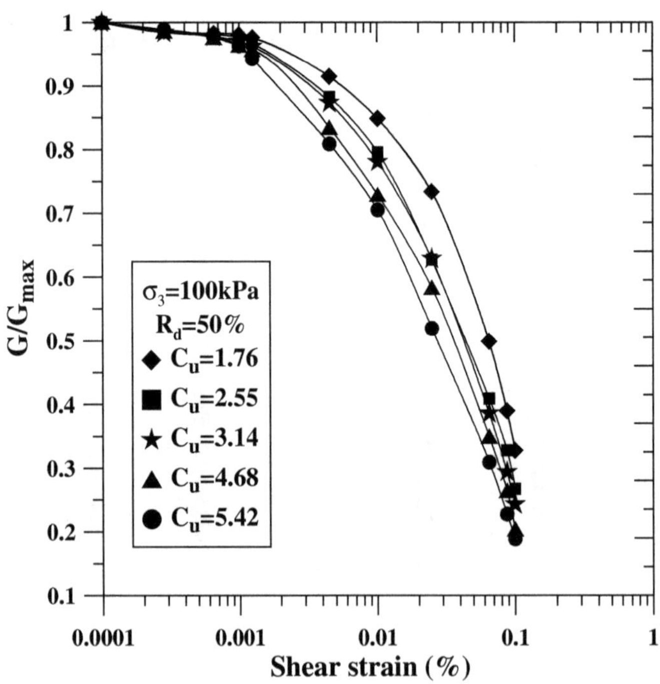

Fig. 5. Variation of G/G_{max} with shear strain for different C_u values

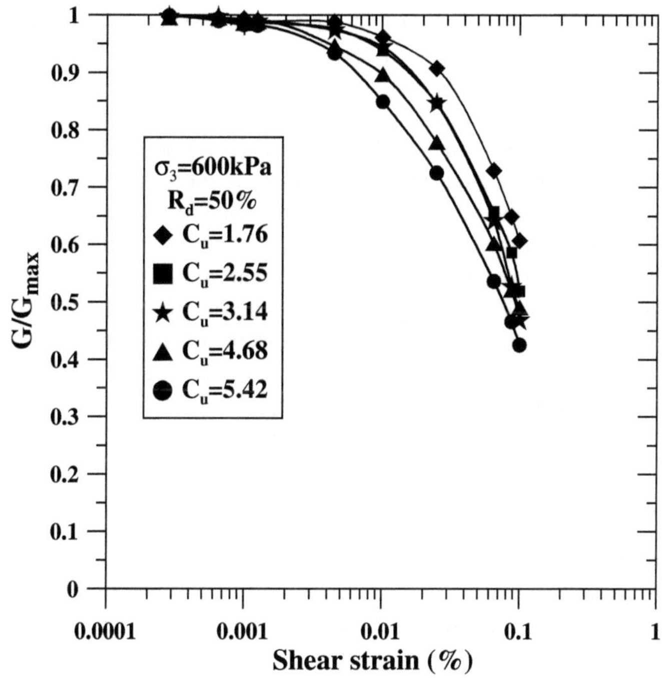

Fig. 6. Variation of G/G_{max} with shear strain for different C_u values

there is a reduction in G/G_{max} from 0.73 to 0.51. similarly, at a higher confining pressure of 600 kPa, shear modulus has reduced from 0.91 to 0.72, as shown in the Fig. 6. This expresses that G/G_{max} also drops with C_u. As discussed, the reason behind this behaviour is variation in inter-particle stress distribution.

Based on the study on force transmission in monodisperse (uniform size particles) and poly-disperse (different size particles) soils by Radjai and Wolf (1998) and Radjai et al. (1998), it is understood that equally distributed force chains are seen in a monodisperse material. In a poly-disperse material, both strong and weak force chains are formed through the inter-particle contacts. As explained in material characterization, the specimens with higher C_u were prepared from different grades of Ennore sand. Hence, soils with higher C_u results in larger stiffness degradation. This behaviour remains same at other confining stresses and shear strains except the variation in the level of influence. At low strains, the difference in G/G_{max} between samples with C_u of 1.76 and 5.42 is very less compared with the difference at the high strain. This is due to induced mobility in the particles at higher strains.

It was also noticed that, in sands, the threshold strain is slightly affected by uniformity coefficient. This threshold shear strain represents the boundaries of the fundamental cyclic behaviour of soil from very small to large strains. This concept of threshold shear strain was first presented by Lo Presti (1991) and later by Vucetic (1994). Research by Radjai and Wolf 1998 on a poly-disperse packing showed that the weak contacts transmit only 28% of the average mean pressure in the granular packing.

It means a quite less number of contacts in a poly-disperse packing have large effective stresses, while larger parts have low γ_t values due to comparably low contact stresses. This led to a lower average γ_t in the poly-disperse packing compared with the monodisperse packing. At a strain below the linear threshold shear strain (γ_t), the soils behave as linear elastic. Between γ_t and volumetric threshold shear strain (γ_v), the soils tend to exhibit recoverable non-linear behaviour, as the microstructure of the soils remains unchanged. Beyond γ_v, the soils behave as highly nonlinear and inelastic due to irreversible changes in the soil microstructure. Based on the past studies, Santos and Correia (2001) proposed a reference threshold shear strain as $\gamma_{0.7}$. This is the shear strain corresponding to a G/G_{max} of 0.7 in the modulus degradation curve. From Figs. 5 and 6, it is seen that with an increase in coefficient of uniformity, the initiation of steepness in the curve was observed at lower shear strains side. In another way, $\gamma_{0.7}$ for a uniformity coefficient of 1.76 is 0.03 whereas, for C_u of 5.42, $\gamma_{0.7}$ is 0.01. This indicates a slight reduction in $\gamma_{0.7}$ with the uniformity coefficient.

5.3 Damping Ratio (D)

Similar to shear modulus, damping is one of the dynamic soil properties which can be defined as a measure of energy dissipation during dynamic loading. In dynamic soil analyses, damping becomes more influential at resonance. Therefore, it would be best if

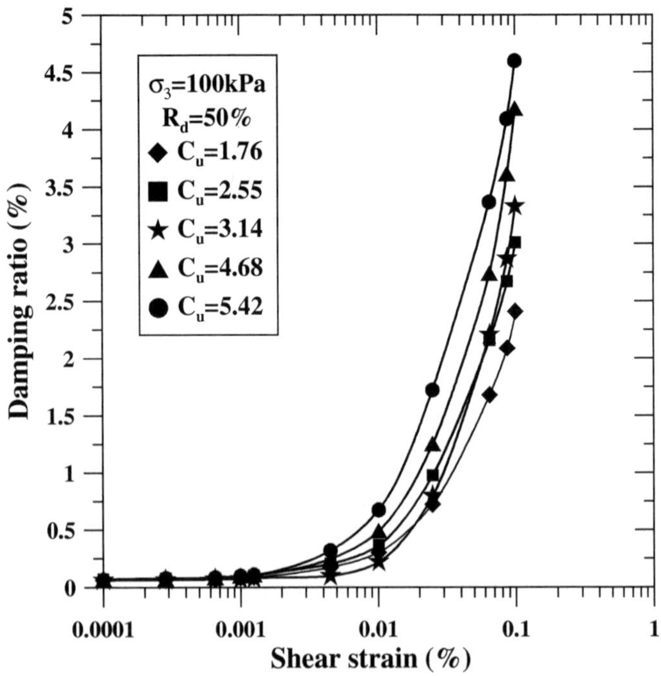

Fig. 7. Variation of damping ratio with shear strain at a confining pressure of 600 kPa for different coefficients of uniformity

laboratory damping measurements are performed in the vicinity of the expected in-situ resonant frequency which would minimize the influence that the selected model has on the overall response (Ashmawy et al. 1995). So, in the present study, the damping properties were also determined.

It was seen from Fig. 7 that, at a shear strain of 0.001%, damping ratio of 0.062%, 0.07%, 0.061%, 0.064%, 0.065% for samples with C_u of 1.76, 2.55, 3.14, 4.68, 5.42, respectively, were perceived. Similarly, at a higher strain level of 0.1%, the observed damping ratio for similar C_u values are 2.4%, 3%, 3.3%, 4.2% and 4.6%. The same behaviour was noticed at a confining pressure of 600 kPa, shown in Fig. 8. This shows that there is negligible effect of uniformity coefficient on damping at small strains, whereas, significant effect is noticed at higher strain levels due to increased mobility. Besides, as the uniformity coefficient increases, damping ratio decreases. This can be attributed to the fact that as the contact area between the particles increases, there will be a rise of space between the particles. So, the energy dispersion will increase and hence, the damping ratio increases.

5.4 Comparison with Data from the Literature

To better understand the effect of C_u, comparison was made between the current test data and data from literature. The comparison of G/G_{max} of the specimens with similar

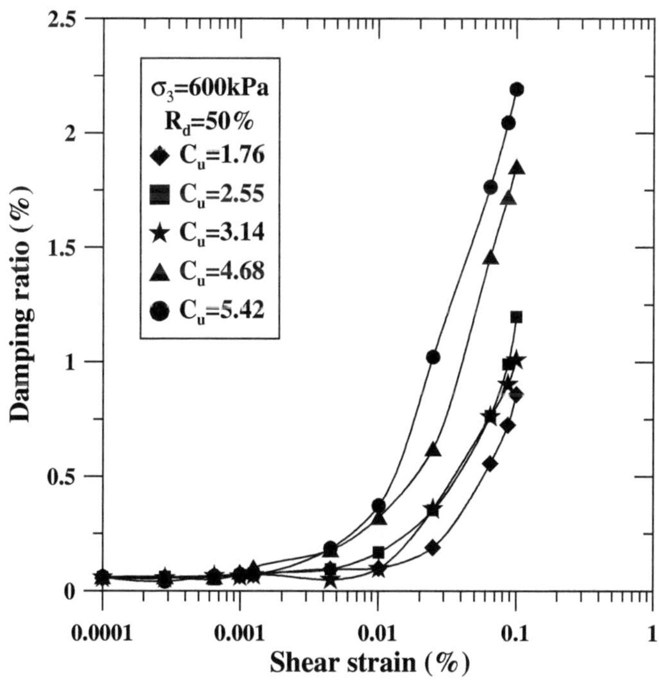

Fig. 8. Variation of damping ratio with shear strain at confining pressure of 600 kPa for different coefficients of uniformity

C_u values from the Wichtmann and Triantafyllidis (2013) study is given in Fig. 9. The Figure shows a reasonable agreement between the present study and Wichtmann and Triantafyllidis (2013) study.

Based on the data produced by Wichtmann and Triantafyllidis (2013) on sands with specially mixed gradations, it was noticed that there is no significant effect of uniformity coefficient on damping ratio. The current test data on sand with similar C_u values also follows a similar trend with C_u, as shown in the Fig. 10 for shear strains of 0.025% and 0.01%. The Figure shows that at a constant confining pressure and relative density, the difference in the damping between the current data and literature at a shear strain of 0.01% is more than the difference at a strain of 0.0025%, from labels C and D of Fig. 10.

The reasons behind this behavioral change are that though both the materials are sands with similar C_u values, the grain size distribution is different. This can be seen from Fig. 1. The range of the particles considered in the Wichtmann study vary from 0.07 mm to 20 mm whereas, in the current study the particles lie in between 0.07 mm to 5 mm. Shape of the soil particles also varies from sub-angular to angular. As the energy dissipation is mainly related to the type of packing, force transmission and particle shape (Payan et al. 2016), the variation is perceived in dynamic soil properties. Usually, with increase in the threshold shear strain, there will be reduction in damping behaviour. From Fig. 2(a–d) of Wichtmann and Triantafyllidis study, it is noticed the

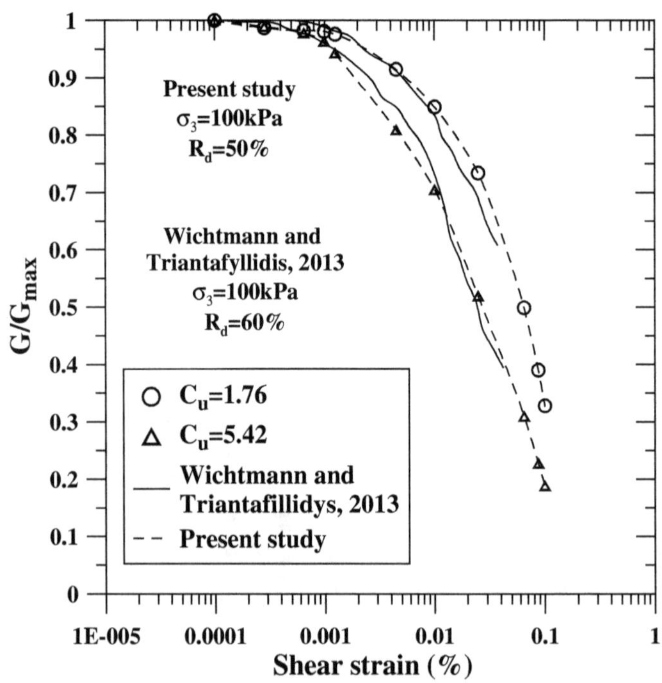

Fig. 9. Comparison of modulus degradation behaviour of present study with Wichtmann and Triantafyllidis (2013) study

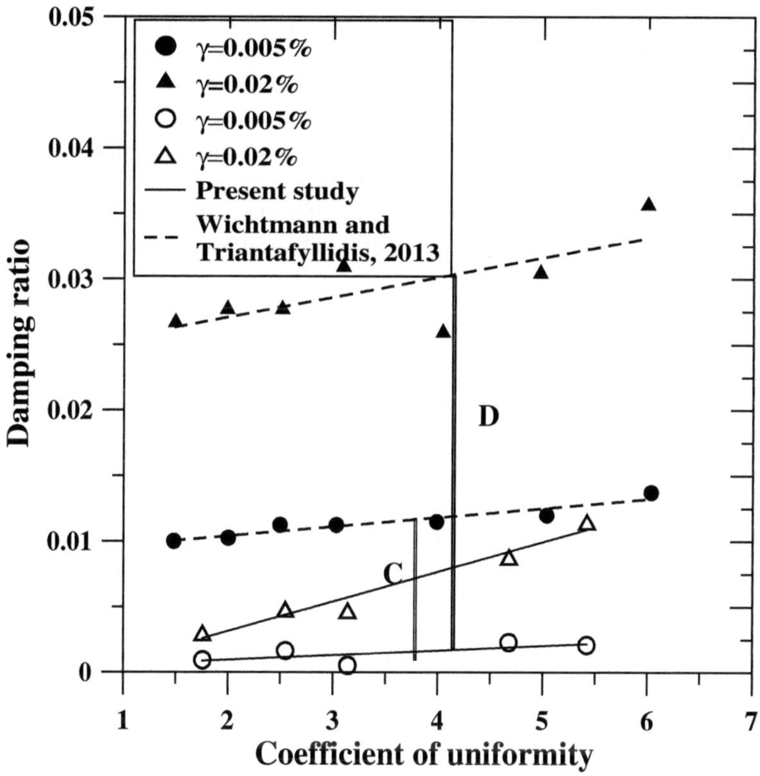

Fig. 10. Comparison of damping ratio behaviour of present study with Wichtmann and Triantafyllidis (2013) study

samples with C_u similar to the present study possess little low $\gamma_{0.7}$. As a result of small $\gamma_{0.7}$, higher maximum particle size and more number of bigger size particles, there is more energy dissipation. Hence, the damping ratio from their study is high compared to the present study.

This shows that the uniformity coefficient significantly influences the modulus degradation behaviour of sands and slightly affects the damping properties. Hence, its effect also needs to be considered in developing any empirical relation for modulus degradation and damping characterization for sands, which mostly remains unique for the soil under consideration.

6 Conclusions

A series of RC tests were performed on Ennore sand specimens with the varied C_u values and the influence of C_u on dynamic properties was studied. The major conclusions drawn are as follows:

- The shear modulus of sand decreases, and damping ratio increases with the increase in the coefficient of uniformity. This is due to uneven stress distribution and increased space between the particles, respectively, with an increase in coefficient of uniformity.
- The increase in the shear modulus and decrease in the damping ratio is almost less in the shear strain range of 0.0001 to 0.001%.
- However, with an increase in uniformity coefficient, there is a little reduction in threshold shear strain portraying that threshold shear strain is slightly affected by gradation (uniformity coefficient).

References

Andrus, R.D., Stokoe, K.H.: Liquefaction resistance of soils from shear-wave velocity. J. Geotech. Geoenviron. Eng. ASCE (2000). https://doi.org/10.1061/(asce)1090-0241(2000)126:11(1015)

Ashmawy, A.K., Salgado, R., Guha, S., Drnevich, V.P.: Soil damping and its use in dynamic analyses. In: The Proceedings of the Third International Conference on Recent Advances in Geotechnical Earthquake Engineering and Soil Dynamics, St. Louis, Missouri, vol. 1, pp. 35–41 (1995)

ASTM D4015-92 (Reapproved 2000) Standard test methods for modulus and damping of soils by the resonant column method

Atkinson, J.H.: Non-linear soil stiffness in routine design. Geotechnique (2000). https://doi.org/10.1680/geot.2000.50.5.487

Dobry, R., Vucetic, M.: State-of-the-art report: dynamic properties and seismic response of soft clay deposits. In: Proceedings of the International Symposium on Geotechnical Engineering of Soft Soils, Mexico City, vol, 2, pp. 51–87 (1987)

Ferreira, C., Martins, J.P., Gomes Correia, A.: Determination of small-strain stiffness of hard soils by means of bender elements and accelerometers. In: Anagnostopoulos, A., et al. (ed.) Proceedings of the 15th European Conference on Soil Mechanics and Geotechnical Engineering, vol. 1, pp. 179–184. IOS Press (2011)

Hardin, B.O., Black, W.L.: Sand stiffness under various triaxial stresses. J. Soil Mech. Found. Div. ASCE. **92**(2), 27–42 (1966)

Hardin, B.O., Drnevich, V.P.: Shear modulus and damping in soils: design equations and curves. J. Soil Mech. Found. Div. ASCE. **98**(7), 667–692 (1972b)

IS 2720 (Part III, Sec 1) – 1980 (reaffirmed 2002), Determination of Specific gravity – fine-grained soils, pp. 1–8

IS 2720 (Part IV) – 1985 (reaffirmed 2006), Grain size analysis, pp. 1–38

IS 2720 (Part XIV) – 1983, Determination of relative density of cohesion less soils, pp. 1–14

Ishibashi, I., Zhang, X.: Unified dynamic shear moduli and damping ratios of sand and clay. Soils Found. **33**(1), 182–191 (1993)

Iwasaki, T., Taksuoka, F.: Effects of grain size and grading on dynamic shear moduli of sands. Soils Found. **17**(3), 19–35 (1977)

Lee, J.S., Santamarina, J.C.: Bender element, performance and signal interpretation. J. Geotech. Geoenviron. Eng. **131**, 1063–1070 (2005)

Lo Presti, D.C.F.: Discussion on threshold strain in soils. In: Proceedings of X ECSMFE, Firenze, Italy, vol. 4, pp. 1282–1283 (1991)

Payan, M., Senetakis, K., Khoshghalb, A., Khalili, N.: Influence of particle shape on small-strain damping ratio of dry sands. Géotechnique (2016). https://doi.org/10.1680/jgeot.15.T.035

Radjai, F., Wolf, D.E., Jean, M., Moreau, J.-J.: Bimodal character of stress transmission in granular packings. Phys. Rev. Lett. **80**(1), 61–64 (1998)

Ravishankar, B.V., Sitharam, T.G., Govinda Raju, L.: Dynamic properties of Ahmedabad sands at large strains. In: Indian Geotechnical Conference, Ahmedabad, India, pp. 369–372 (2005)

Santos, J.A., Correia, A.G.: Reference threshold shear strain of soil, its application to obtain a unique strain-dependent shear modulus curve for soil. In: Balkema, A.A. Proceedings of the 15th International Conference on Soil Mechanics and Geotechnical Engineering, Istanbul, vol. 1, pp. 267–270 (2001)

Silver, M.L., Seed, H.B.: Deformation characteristics of sands under cyclic loading. J. Soil Mech. Found. Div. ASCE **97**(8), 1081–1088 (1971)

Viana da Fonseca, A., Ferreira, C., Fahey, M.: A framework for interpreting bender element tests, combining time-domain and frequency-domain methods. Geotech. Test. J. ASTM **32**(2), 1–17 (2009)

Vucetic, M.: Cyclic threshold shear strains in soils. J. Geotech. Eng. **120**(12), 2208–2228 (1994)

Wichtmann, T., Triantafyllidis, T.: Effect of uniformity coefficient on G/G_{max} and damping ratio of uniform to well-graded quartz sands. J. Geotech. Geoenviron. Eng. 59–72 (2013) https://doi.org/10.1061/(ASCE)GT.1943-5606.0000735

Yang, J., Yan, X.R.: Site response to multidirectional earthquake loading: a practical procedure. Soil Dyn. Earthq. Eng. **29**, 710–721 (2009)

Zen, K., Umehara, Y., Hanada, K.: Laboratory tests and in-situ seismic survey on vibratory shear modulus of clayey soils with different plasticities. In: Proceedings of 5th Japan Earthquake Engineering Symposium, Tokyo, pp. 721–728 (1978)

Dynamic Properties of Surface Liquefied Site Silty-Sand of Tripura, India

Kunjari Mog(✉) and P. Anbazhagan

Department of Civil Engineering, Indian Institute of Science, Bangalore 560012, India
mogkunjari@gmail.com, anbazhagan@iisc.ac.in

Abstract. Due to increase in the rate of occurrence of earthquakes and associated damages, site-specific dynamic models are attracted to research in the recent past. The shear modulus degradation curve, damping characteristics and shear wave velocity profile are indispensable input properties required for carrying out the site response analysis. However, most of the site response studies are being carried out using existing dynamic model developed for generic soil. In the present study, results of resonant column and cyclic triaxial tests on reconstituted dry silty-sands specimens obtained from the recently liquefied site of Tripura, India is presented. The shear modulus degradation curve and damping ratio for a low ($<10^{-3}$) to high strain range (up to 5%) are presented and compared with the available well-known curves and discussed. Also, the effects of confining pressure and relative density on dynamic properties are presented. Further, comparison has been made between the shear wave velocity measured in the liquefied field using Multichannel Analysis of Surface Waves (MASW) and laboratory results of low strain shear wave velocity. These results can be used for amplification estimation or micro-zonation studies in Tripura.

Keywords: Shear modulus · Damping · Silty-sands · Resonant column
Cyclic triaxial · In-situ shear wave velocity

1 Introduction

Tripura state is in the north-eastern part of India and it is close to the Himalayan belt that is seismically very active due to the convergent boundary of Indian plate with Eurasian plate. The state falls in seismic zone V as per the Indian seismic code (IS: 1893 2016) with an anticipated zero period acceleration of 0.36 g. Recently, Tripura is jolted by a moderate magnitude of 5.7 (Mw) in the year 2017, January 3rd (USGS; Debbarma et al. 2017; Anbazhagan et al. 2017). The epicenter was traced at Ambassa, Dhalai district of Tripura, India. It brought the liquefaction at agricultural farm land that caused extensive damages to crop and developed numerous cracks in a number of dwellings. It also led the lateral spreading in the river bank at Kanchanbari, Kumarghat area. The Geophysics Division, NER, Shillong, reported that soil type in Tripura is generally consists of unconsolidated sediments of thickness about more than 250 m, comprise mainly sand and silty soils. The large variations of depth to bedrock and

D. Choudhury et al. (Eds.): GeoMEast 2018, SUCI, pp. 140–154, 2019.
https://doi.org/10.1007/978-3-030-01920-4_12

high-water table along with presence of soft silty sands dictate the preparation of seismic hazard maps, dynamic properties and liquefaction studies in Tripura.

Since early 1970's numerous significant researches come out on dynamic soil properties of soil by several researchers (Seed and Idriss 1970; Hardin and Drnevich 1972; Kokusho 1980; Woods 1991; Vucetic and Dobry 1991; Ishibashi and Zhang 1993; Zhang et al. 2005). Numerous model curves have been developed on shear modulus either on sands or clays although limited amount of work have been reported on gravelly soils or silty sands. The site-specific laboratory dynamic soil properties such as shear modulus and damping ratio are widely recognized as indispensible input parameters for carrying out the site-specific response analysis, liquefaction assessment, soil structure interaction, and seismic slope stability analysis. Similarly, geo-physical methods have become popular to geotechnical engineers to estimate the in-situ shear wave velocity (V_s) which in turns is used for site classification of soils, estimating the dynamic soil properties, liquefaction assessment, mapping the stratigraphic soil layers, detecting cavities or sinkholes and micro zonation studies or estimating the site-specific amplification estimation. The variation of V_s is dependent on soil density, soil type, effective stress, void ratio, depositional environment, etc. Most often the correlation between the Vs and SPT N values are used in places where geophysical measurement is not feasible or where the geophysical data is not available.

In the present study, dynamic properties and model of silty-sands is reported which had been emitted to the ground surface during 3rd January 2017 Tripura earthquake. The resonant column and cyclic triaxial tests have been conducted on the reconstituted soil samples of size 50 mm by 102 mm. The shear modulus degradation curve and damping ratio for a low ($<10^{-3}$) to high strain range (beyond 5%) are presented and compared with the available well-known curves and discussed. Further, the shear wave velocity measured in the liquefied field using Multichannel Analysis of Surface Waves (MASW) is compared with the laboratory (low strain) shear wave velocity.

2 Shear Modulus Degradation and Damping Curves

The experimental investigation was carried out on the surface liquefied soil obtained from the Kanchanbari, Tripura, India. Tripura state has experienced several moderate to big earthquakes in the past. Recently, an earthquake of magnitude Mw 5.7 hit the state on 3rd January 2017 which tremor was felt nearly all the north-eastern states of India including neighbouring country Bangladesh and Myanmar (Anbazhagan et al. 2017). Though the earthquake was of moderate magnitude, it caused the liquefaction in agricultural farmland in Kanchanbari, Tripura. It caused extensive damages to crop and developed numerous cracks in a number of houses. It also caused the lateral spreading in the river bank at Kanchanbari, Kumarghat area. The large variations of depth to bedrock and high-water table along with the presence of soft silty sands necessitate the preparation of seismic hazard maps, dynamic properties and liquefaction studies in Tripura.

The soil samples obtained from the site was sieved carefully and the grain size distribution curve is shown in Fig. 1. The basic property of soil was determined through laboratory test that is summarized in Table 1 (IS: 2720). Based on the Indian

Soil Classification System (IS: 2720), the soil can be classified as silty-sand. The undisturbed sample has been collected in a thin Shelby-tube sampler from a 3 m depth to measure the in-situ density and it is measured as 1.59 g/cc. The minimum and maximum relative density determined in the lab is 1.741 g/cc and 1.394 g/cc. The shear modulus degradation and damping curves for a low to high strain ranges are measured and curve has been generated. Further the effect of confining pressure and relative density on dynamic soil properties are discussed.

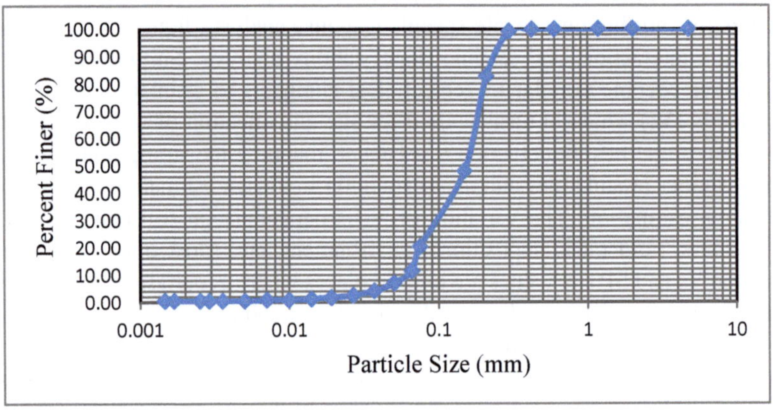

Fig. 1. Grain size distribution of tested soil

Table 1. Index properties of soil sample

Property	Value
Specific gravity, (G_s)	2.636
Particle size distribution	
(a) Gravel	0
(b) Fine Sand	79.753
(c) Silt & Clay	20.247
(d) Silt	19.837
(e) Clay	0.410
Uniformity coefficient (C_u)	2.903
Coefficient of Curvature (C_c)	0.861
Plastic limit	Non-plastic
Maximum dry unit weight, γ_{dmax}	1.741 g/cc
Minimum dry unit weight, γ_{dmin}	1.395 g/cc
Insitu dry density	1.66 g/cc
Indian standard classification	Silty-sand

2.1 Sample Preparation

A constant volume split mould of size 50.0 mm internal diameter and 102 mm height was used in preparing the test specimens. All the test specimens were prepared with a relative density of 80%, 50% and 30% using the dry tamping method reported by Ladd (1978). The same specimen preparation technique was adopted for both the resonant column and cyclic triaxial test. At first, an appropriate amount of air-dried representative sand was considered for preparing the test specimens. A latex rubber membrane was fixed to the bottom platen by securing it with O-rings. Next, the split mould was placed over the bottom platen, outside of the rubber membrane. The top part of the rubber membrane was strained and folded over the split mould. A constant vacuum of 15 to 30 kPa was applied through the bottom platen in order to remove the entrapped air between the membrane and the mould. The advantage of applying vacuum is that the membrane firmly sticks to the mould and it makes the sample stable during the preparation stage; also, it helps to reduce the sample disturbance while removing the mould. The sand was then gently poured into the split mould from a free drop height of 15 to 40 mm. The samples were prepared in six equal thick layers. Each layer was compacted using the tamping rod which weighs 142 g. In order to maintain a uniform density throughout the specimen height, the number of blows given to the lower layer was less compared to the successive upper layers. The volume of soil placed in each layer nearly corresponds to a 17 mm thick and the height of 102 mm was maintained varying the number of blows.

2.2 Resonant Column Test

Resonant Column (RC) instrument has been used widely in the laboratory for measuring the dynamic soil properties at low to medium strains (in either undisturbed or reconstituted states) as early as 1930's. Since then many different designed for resonant column devices have been developed and implemented by several researchers (Hardin and Richart 1963; Drnevich and Richart 1970; Stokoe et al. 1994) throughout the world. Three types of resonant column apparatus commonly available are fixed-fixed type, free-free and fixed-free type, based on the boundary conditions.

A Fixed-free GCTS resonant column device has been used in this study, where the soil column is fixed at the base and free to rotate at the top. A cylindrical soil specimen is excited at the top by an electromagnetic loading system or motor. A harmonic torsional excitation over a wide range of frequencies is applied at the top of the specimen to vibrate in one of its natural modes of vibration and the frequency response curve is drawn. The frequency, at which the amplitude is maximum, delineates the resonance frequency of the specimen. Once the resonance frequency is known, the shear wave velocity and corresponding shear modulus can be easily determined using the elasticity theory. The equation used to compute the maximum shear modulus is as follows:

$$G_{max} = \rho * V_{s^2} \tag{1}$$

Where: ρ = Density, V_s = Low strain shear wave velocity.

Initially, lower amplitude of torque was applied through a motor to the specimen and then the amplitude was gradually increased until the desired strain level is attained (0.01%). In the GCTS resonant column apparatus, the amplitude of the drive system or motor is applied in terms of percentage full scale (pfs) value that correlates the maximum output voltage. In addition, damping value is obtained from the recording of the free-vibration decay after the electromagnetic drive system was shut-off.

The resonant column tests were conducted on reconstituted cylindrical soil specimen of size 50 mm in diameter and 102 mm height. All the tests were conducted on the dry specimen. The relative density of 80%, 50% and 30% were considered for the resonant column test and the results obtained for the confining pressures of 50, 100, and 200 kPa are presented in the proceeding sections.

2.3 Cyclic Triaxial Test

After resonant column test is completed for the desired strain amplitude of 0.01%, the cell pressure from the triaxial chamber was released slowly and simultaneously the vacuum pressure of 15–20 kPa were applied to the soil specimen to keep the sample stand upright. Although small strains generated during the resonant column test would result in some rearrangement of the grains of soils in the dry specimen, however, it was found that the dry density has changed from the initial 1.48 g/cc to final 1.49 g/cc after the resonant column test as measured for 30% relative density test specimen and no change was observed for 80% and 50% relative density. Hence, changes in the relative density measured after the resonant column test remains nearly the same for 80% and 50% tests specimen except slight variation for the 30% test specimen. Therefore, keeping all the conditions (such as confining pressures, loading frequency etc.) same, next, cyclic triaxial tests were performed on the same specimen.

The instrument used for the cyclic triaxial test is servo-controlled that equipped with a load capacity of 10 kN pneumatic actuator. The system can perform both the strain controlled and stress-controlled tests with a frequency range of 0.01 Hz to 10 Hz. The different type of waveforms such as sine, triangular, square or any other type of random loading waveforms can be employed as an external input in the system. For the present study, the strain-controlled cyclic triaxial test was performed to determine the dynamic soil properties at medium to large strains. The double amplitude sinusoidal loading was applied for all cyclic triaxial tests at a loading frequency of 1 Hz. The load was applied through a low friction loading piston, the upper end to which a servo-controlled pneumatic actuator is fitted and the lower end is rigidly connected to the top of the specimen. The measurement of axial deformation, cell pressure, and pore water pressure for saturated specimen etc. can be made using a built-in data acquisition system. The output of cyclic triaxial tests is the plot of deviator stress (σ_d), versus axial strain (ε), also known as the hysteresis loop. A typical hysteresis loop for one cycle of loading is shown in Fig. 2 in which the slope of the secant line joining the extreme point is the dynamic Young's modulus, E. In the cyclic triaxial test, initially E is measured using $E = \sigma_d/\varepsilon$ and then shear modulus (G), shear strain (γ) are calculated from the following relationship:

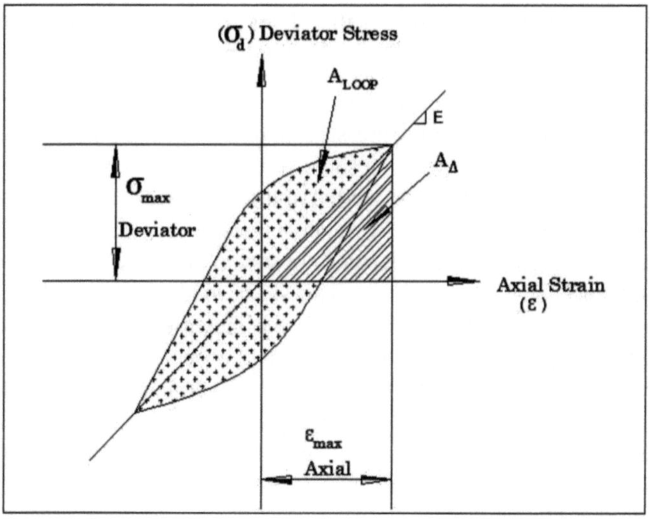

Fig. 2. Hysteretic stress-strain behaviour for one cycle of loading

$$G = E/2(1+\mu) \quad \text{and} \quad \gamma = (1+\mu)\varepsilon \tag{2}$$

Where: μ = Poisson's ratio. The μ value may be assumed as 0.4 for dry sand specimens as recommended by Silver and Park (1975).

The damping ratio D is a measure of the dissipated energy to the elastic strain energy (Kramer 1996), and it can be estimated from the area of hysteresis loop using the given equation:

$$D = \frac{A_{loop}}{4\pi A_{\Delta}} \tag{3}$$

Where, A_{loop} = Area enclosed by the hysteresis loop and A_{Δ} = Area of the triangle. Dynamic soil properties of large shear strains were obtained using a series of strain-controlled cyclic triaxial. The damping properties using both the conventional formula as given in Eq. 3 and modified asymmetrical hysteresis loop method as suggested by Kumar et al. (2017) as given in Eq. 4 is calculated and the variation between the two was found to be approximately 20 to 40% for the tested silty-sand samples. However, the damping ratio calculated using the Eq. 3 is reported in the present study. The modified damping estimation equation considering the asymmetrical hysteresis loop is given below and more details can be found in Kumar et al. (2017):

$$D^{\#} = \frac{A_{loop}}{\pi(A_{\Delta 1} + A_{\Delta 2} + A_{\square})} \tag{4}$$

Where $D^{\#}$ = Damping ratio by modified method, A_{loop} = Area enclosed by the hysteresis loop and $A_{\Delta 1} + A_{\Delta 2}$ = Area of the two triangles, $A_{\square}$ = Area of rectangle.

2.4 Effect of Confining Pressure on the Shear Modulus and Damping Properties at Large Shear Strain (Beyond 5%)

The resonant column and cyclic triaxial (strain controlled) tests were carried out to investigate the influence of confining pressure on dynamic properties of dry silty-sands. The variation of shear modulus with the shear strain at low ($<10^{-3}$) to high strain range (more than 5%) for the dry silty-sands tested at three different confining pressures (50, 100, 200 kPa) are shown in Fig. 3. The relative density considered for testing the soil specimen was 80%. From the Fig. 3 it can be observed that the confining pressure has a significant influence on shear modulus. The shear modulus (G) decreases with the increase in shear strain and G increases with the increase in confining stress. A noticeable variation in the value of shear modulus was observed for all the three different confining pressures (strain range 0.0001 to 1%), however, beyond 1% of shear strain, it follows the converging trends regardless of any confining pressures to which the soil specimens are subjected. The similar converging trends of the shear modulus (beyond 1%) has also been reported in the literature (Sitharam et al. 2008).

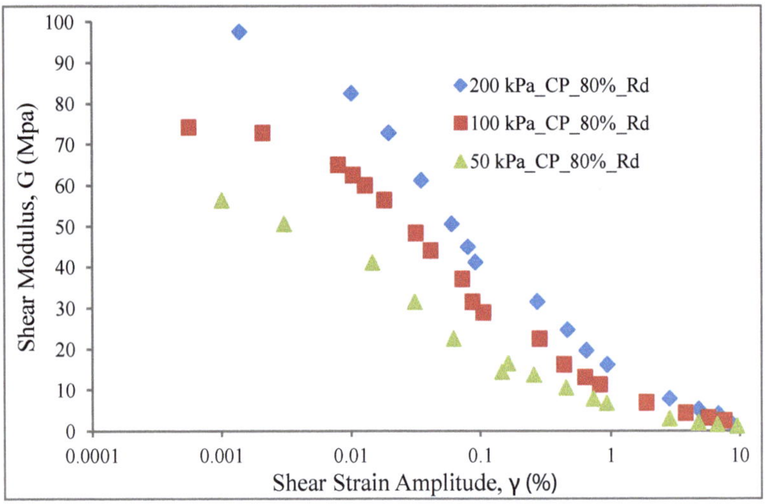

Fig. 3. Effect of confining pressures on shear modulus for a large shear strain

The influence of confining pressures on damping ratio is shown in Fig. 4. It can be observed from the figure that the damping ratio increases with the increase in shear strain amplitude for a given confining pressure. The damping ratio is observed to be maximum for a lower confining pressure (50 kPa) compared to that of higher confinement (100 kPa and 200 kPa). It can also be seen that the damping ratio is nearly equal up to a shear strain of 0.01%, however, beyond that strain level damping ratio decreases with the increase in confining pressures. It is also evident from the Fig. 4 that up to a shear strain of about 2%, the damping shows the conventional behavior of increasing trend with increasing shear strains, however, the decreasing trends of damping is noticeable beyond 2% of shear strain for all the three different confining pressures. This decrement of damping at higher strains differs from the conventional trends (Damping increases or became asymptotic with increasing shear strains, beyond 1%) and is concords with the findings of Kumar et al. (2017).

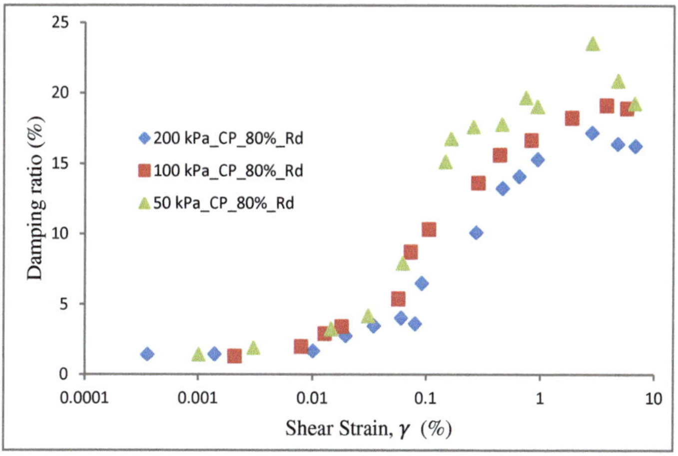

Fig. 4. Effect of confining pressures on damping ratio for a large shear strain

It is to be noted here that most of the damping curves reported in the literature are for the shear strain of less than 1% (Seed and Idriss 1970; EPRI 1993; Roblee and Chiou 2004) and beyond 1% of strain level the damping value is extrapolated or predicted (Kumar et al. 2017). The existing commercial software for dynamic analysis such as FLAC, SHAKE, and DEEPSOIL etc. also uses the same extrapolated conventional damping ratio curve. Hence, from the present study, it can be stated that conventional trends of damping ratio curve may not be appropriate to use in the geotechnical analysis for a wide strain range and it should consider the decreasing trends of damping behavior at large shear strain as shown in the Fig. 4.

2.5 Effect of Relative Density on Shear Modulus and Damping at Large Strain

The effect of relative density on shear modulus for a wide range of shear strain was investigated by performing RC and Cyclic Triaxial tests at three different relative densities such as 80%, 50%, and 30% respectively. The shear modulus variations with the shear strain with different relative densities are shown in Fig. 5. The result is shown for the confining pressure of 200 kPa. It is evident from the Fig. 5 that the shear modulus decreases with the decrease in relative density (80%, 50%, and 30%) up to a strain level of 0.5% and beyond that no variation in shear modulus is found. Hence, this clearly indicates that relative density has considerable influence on small strain shear modulus of silty-sands (less than 0.5%); however, it does not have an influence on large strain shear modulus at 200 kPa confining pressure.

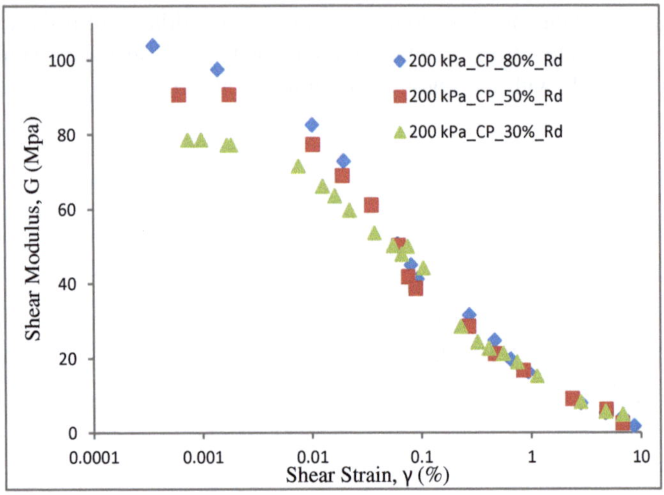

Fig. 5. Effect of relative density on shear modulus at 200 kPa confining pressure

Figure 6 shows the effect of relative density on damping ratio for a wide strain range. The dry silty-sand specimen was subjected to 200 kPa confining pressure for the relative density of 80%, 50% and 30% respectively. From the Fig. 6 it can be clearly observed that the relative density has negligible or no influences on damping ratio at various strain levels (Sitharam et al. 2008); only a slight variation in damping ratio is observed beyond 2% of shear strain in the present study.

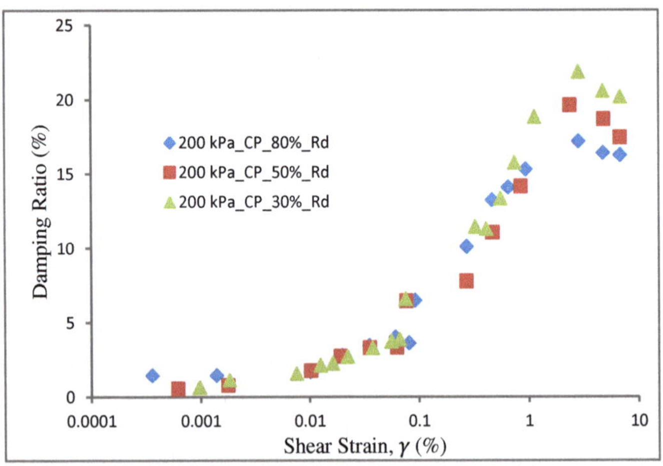

Fig. 6. Effect of relative density on damping ratio at 200 kPa confining pressure

2.6 Normalized Shear Modulus and Damping Curves

It is a general trend to represent the variation of shear modulus with respect to shear strain in terms of normalized shear modulus degradation curves. This normalized curve is then used for site response analysis or amplification estimation. The normalized shear modulus degradation curve for the silty-sand tested for different relative density (80%, 50%, 30%) for different confining pressures (200, 100, 50 kPa) is presented in Fig. 7 and the results is compared with the available well-known curves of sand. It can be observed from the Fig. 7 that the shear modulus obtained from the combined resonant column and cyclic tri-axial tests for silty-sands do not fit well within the upper and lower boundary range as proposed by Seed and Idriss (1970) for sands. However, the curvilinear shape or trend of normalized (G/G_{max}) curve is same as that of sand. Further, this normalized curve is compared with the other available curves for sands (EPRI 1993; Roblee and Chiou 2004) and shown in Fig. 7. The data presented in the present study exceeds the upper limit of Seed & Idriss curve and so as the case for other available curves as mentioned. This clearly brings about the fact that the presence of fines content caused the cyclic shearing strength to increase and the shear modulus of silty-sand show the higher value than the shear modulus of sands.

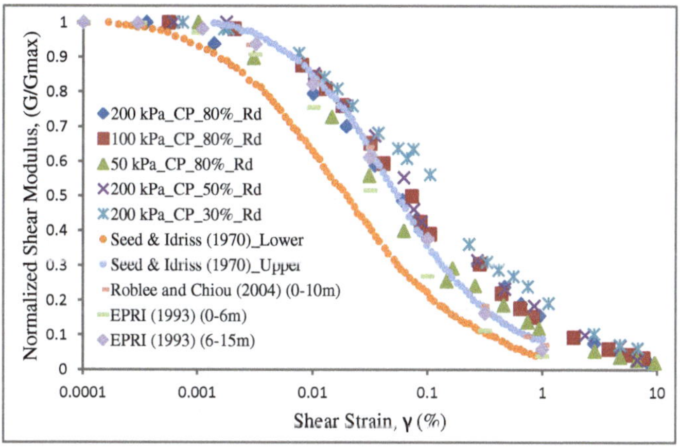

Fig. 7. Normalized shear modulus curve of silty-sands obtained from combined resonant column and cyclic triaxial test compared with sands

Also, as can be seen from the Fig. 8 that the range of damping for the present study falls much lower side from the lower bound curve of proposed Seed and Idriss (1970), EPRI (1993), Roblee and Chiou (2004) curve for sands. Based on the observed data, it can be concluded that strain-dependent shear modulus ratio of silty-sands is higher than that of sands and damping properties is lower than sands. The site specific dynamic shear modulus reduction curves developed from the present study could be used for site response studies or amplification estimation in the area. However, as the damping ratio curves for dry and saturated cohesionless soils are not the same at high strains

(Kumar et al. 2017), hence, the damping ratio curves as reported in the present study may be further studied for statured condition and can be used for the dynamic analysis.

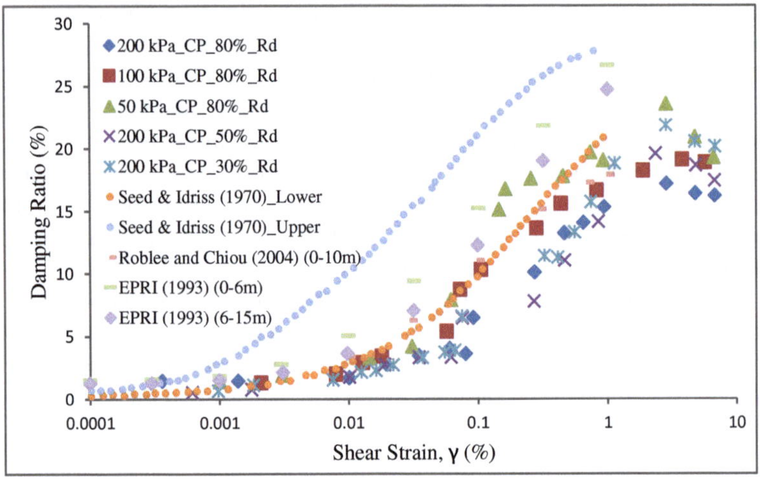

Fig. 8. Damping curve of silty-sands obtained from combined resonant column and cyclic triaxial test compared with sands

3 Field Measurement of V_s and Comparison with Laboratory V_s

The measurement of in-situ shear wave velocity and the comparison between the in-situ V_s and laboratory V_s is reported in this section. It is widely recognized fact that shear wave velocity profile is fundamental input parameter required in micro-zonation studies or in estimating the site-dependent amplification factor. The five sets of Multichannel Analysis of Surface Wave (MASW) tests were conducted at the liquefied site to measure the V_s and the average V_s has been considered to compare with laboratory V_s.

3.1 In-situ Measurement of Shear Wave Velocity

The geophysical methods have become popular among the geotechnical engineers and researchers to estimate the in-situ shear wave velocity (V_S) in the earthquake-prone areas. Various methods presently available to measure the V_s includes Spectral Analysis of Surface Wave (SASW) tests, Multichannel Analysis of Surface Wave (MASW) Tests, suspension logging tests, cross-hole or down-hole tests (Kramer 1996). Both the cross-hole and down-hole methods use body waves to estimate the shear wave velocity and data obtained from these measurements are fairly accurate. Though the cross-hole test requires drilling of minimum two boreholes and down-hole or up-hole test can be performed in a single boring, often the cost and time associated with these methods are huge; also, drilling of boreholes in urban areas often needs approval from the concerned authority and often drilling is not feasible in some areas. To alternate, these destructive methods, non-destructive SASW or MASW test is preferred. The SASW and MASW test is the well-recognized technique which uses the Rayleigh waves for estimating the in-situ shear wave velocity.

In the present study, the MASW test has been conducted at the liquefied site of Kanchanbari, Tripura, India. The test method involves field testing and data processing which includes dispersion estimation and inversion. First, the 24 geophones with a minimum frequency of 2 Hz are placed in a linear arrangement at an equal interval of 2 m spacing. Then all the geophones are linked to the multichannel recorder. Next, a sledgehammer of 7.8 kg is dropped at one end of geophone line from a certain height to hit the 300 mm × 300 mm metal plate having 25 mm thickness which produces the vibration at the ground. Due to this vibration, the surface wave travels from the first geophone to the last geophone and data is recorded. The SurfSeis software has been used for processing the recorded data. The results obtained from the MASW test is presented in terms of variation of shear wave velocity with depth as shown in Fig. 9. It can be observed from the Fig. 9 that the shear wave velocity is in the range of 88 m/s to 228 m/s for the depth of 0.5 m to 12.5 m, which indicates that the stiffness beneath the ground surface at the site is soft or medium.

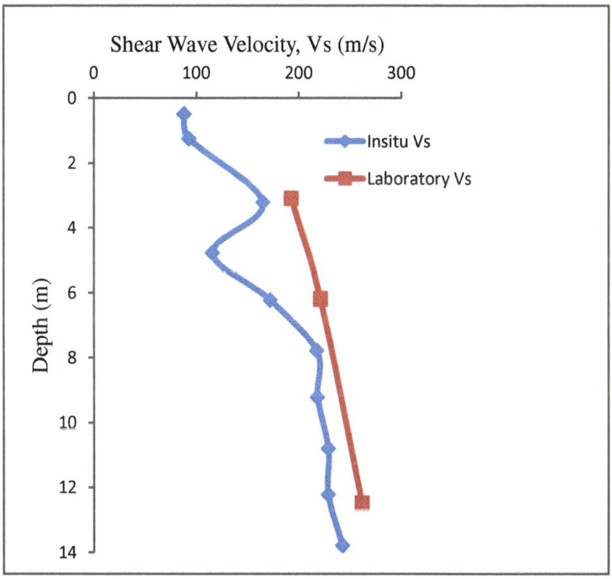

Fig. 9. Comparison of in-situ and laboratory shear wave velocity with depth

To compare the in-situ V_s with laboratory V_s, the reconstituted soil specimen prepared at 80% relative density which is close to the in-situ dry density (1.66 g/cc) at the site has been used and the result is shown in Fig. 9. The in-situ dry density was measured by extracting the partially undisturbed tube sample obtained from a depth of

3 to 3.5 m that was directly brought to the lab from the liquefied site. It can be observed from Fig. 9 that the low patches of shear wave velocities of depths at less than 7 m (or 3.5 to 5 m) is indicative of the possible in-situ liquefaction zones. However, it is amply clear that the laboratory results at these depths are noticeably higher than in-situ observation; the reason being once the site gets liquefied, and the pore-water is dissipated, a significant rearrangement of the cohesionless soil grains occurs, which generates a different void ratio, relative density and stress-state of the soil. Under such condition, it is primarily a different soil state than that present before liquefaction. The small-strain shear modulus, Gmax, and thus, Vs, depends largely on the above-mentioned parameters (Kramer 1996; Ishibashi and Zhang 1993). It can also be seen that the stress state beyond 8 m depth is noticeably same possibly due to the non-liquefied medium, and hence, the results are in good agreement from depths 8 m–12.5 m. Though the V_s profile is available for the deeper depth, however, since the laboratory tests have been conducted up to a confining pressure of 200 kPa which represents the 12.5 m depth from the ground surface, the comparison between the two has been shown up to the same depth.

4 Conclusions

The paper presented the development of site-specific shear modulus and damping ratio curved through the laboratory and in-situ experimental investigation. Dry silty-sand samples obtained from the liquefied site of Trippura, India were subjected to resonant column and cyclic triaxial tests to develop the dynamic properties over varying strain range. MASW tests have also been conducted at the field to obtain the variation of shear wave velocity with depth, and comparative results of the laboratory and in-situ tests have been provided.

The shear modulus of the silty-sand is found to be affected significantly by the variations of confining pressures (at small to large strain range) and relative density (at small strains). However, the damping ratio exhibits no influence over the variations of the relative density. The damping properties show the conventional behavior of increasing trend with increasing shear strains; however, the decreasing trends of damping is observed beyond 2% of shear strain for all the confining pressures to which the samples are subjected. This reduction of damping ratio at higher strains (beyond 2%) differs from the conventional extrapolated trends and should be carefully considered while using the site response analysis or other dynamic analysis.

The normalized shear modulus ratio compared with the available well-known curves for sands revealed the fact that silty-sands exhibits higher modulus ratio than that of sands and exactly opposite trends is observed in case of damping ratio. Based on the present study it can also be interpreted that the possible liquefied layer was at the depth of 3.5 to 5 m for the liquefied site of Kanchanbari, Tripura, India, which has the V_s value less than 150 m/s.

Acknowledgment. The authors thank the Board of Research in Nuclear Sciences (BRNS) of the Department of Atomic Energy (Dae), Government of India for funding the project titled "Probabilistic seismic hazard analysis of Vizag and Tarapur considering regional uncertainties", Ref: BRNS/36016-2016.

References

Anbazhagan, P., Sitharam, T.G.: Evaluation of dynamic properties and ground profiles using MASW: correlation between Vs and N 60. In: Proceedings of 13th Symposium on Earthquake Engineering, Indian Institute of Technology, Roorkee, December 2006, pp. 18–20 (2006)

Anbazhagan, P., Mog, K., Reddy, G.R.: Liquefaction in Kanchanbari Tripura, India due to Moderate Magnitude Earthquake. Current Science (Under review) (2017)

Debbarma, J., Martin, S.S., Suresh, G., Ahsan, A., Gahalaut, V.K.: Preliminary observations from the 3 January 2017, MW 5.6 Manu, Tripura (India) earthquake. J. Asian Earth Sci. **148**, 173–180 (2017)

GCTS-CATS: Resonant column & torsional shear test. User Guide & Ref 1.97:1-143 (2007)

Seed, H.B., Idriss, I.M.: Soil Moduli and Damping Factors for Dynamic Response Analyses. Report No. EERC 70-10, University of California, Berkeley, December 1970. Also Section 5, Soil Behavior Under Earthquake Loading Conditions, Report to U.S. Atomic Energy Commission, SW-AA (1972)

Hanumantharao, C., Ramana, G.V.: Dynamic soil properties for microzonation of Delhi, India. J. Earth Syst. Sci. **117**(2), 719–730 (2008)

Hardin, B.O., Drnevich, V.P.: Shear modulus and damping in soils: design equations and curves. J. Soil Mech. Found. Div. **98**(sm7) (1972)

Imai, T., Yoshimura, M.: The relation of mechanical properties of soils to P and S wave velocities for soil ground in Japan. Urana Research Institute, OYO Corp. (1972)

IS: 1893 (Part 1): Criteria for earthquake resistant design of structures – Part 1: General provisions and buildings; Bureau of Indian Standards, New Delhi (2016)

IS: 2720 (Part 14): Methods of test for soils: Determination of density index (Relative Density) of cohesionless soils. Bureau of Indian Standards (BIS), New Delhi (1986)

IS: 2720 (Part 3): Methods of test for soils: determination of specific gravity. Bureau of Indian Standards (BIS), New Delhi (1980)

IS: 2720 (Part 4): Methods of test for soils: Grain size analysis. Bureau of Indian Standards (BIS), New Delhi (1983)

Ishibashi, I., Zhang, X.: Unified dynamic shear moduli and damping ratios of sand and clay. Soils Found. **33**(1), 182–191 (1993)

Jafari, M.K., Shafiee, A., Razmkhah, A.: Dynamic properties of fine grained soils in south of Tehran. J. Seismol. Earthq. Eng. **4**(1), 25 (2002)

Kokusho, T.: Cyclic triaxial test of dynamic soil properties for wide strain range. Soils Found. **20** (2), 45–60 (1980)

Kramer, S.: L.(1996). Geotechnical Earthquake Engineering. Pren-tice Hall, New Jersey (2005)

Kumar, S.S., Krishna, A.M., Dey, A.: Evaluation of dynamic properties of sandy soil at high cyclic strains. Soil Dyn. Earthq. Eng. **99**, 157–167 (2017)

Ladd, R.S.: Preparing test specimens using undercompaction. Geotech. Test. J. **1**(1), 16–23 (1978)

Maheshwari, B.K., Mahajan, A.K., Sharma, M.L., Paul, D.K., Kaynia, A.M., Lindholm, C.: Relationship between shear velocity and SPT resistance for sandy soils in the Ganga basin. Int. J. Geotech. Eng. **7**(1), 63–70 (2013)
Rollins, K.M., Diehl, N.B., Weaver, T.J.: Implications of Vs-BPT (N 1) 6 0 correlations for liquefaction assessment in gravels. In: Geotechnical Earthquake Engineering and Soil Dynamics III, pp. 506–517. ASCE (1998)
Silver, M.L., Park, T.K.: Testing procedure effects on dynamic soil behavior. J. Geotech. Geoenviron. Eng. **101**(ASCE# 11671 Proceeding) (1975)
Sitharam, T.G., Ravishankar, B.V., Vinod, J.S.: Dynamic properties of dry sands. Indian Geotech. J. **38**(3), 334–344 (2008)
Sykora, D.W., Koester, J.P.: Correlations between dynamic shear resistance and standard penetration resistance in soils. In: Earthquake Engineering and Soil Dynamics II—Recent Advances in Ground-Motion Evaluation, pp. 389–404. ASCE, June 1988
United State Geological Survey (USGS), 3rd January 2017, M 5.7 – 20 km ENE of Ambassa, India
Vucetic, M., Dobry, R.: Effect of soil plasticity on cyclic response. J. Geotech. Eng. **117**(1), 89–107 (1991)
Woods, R.D.: Field and laboratory determination of soil properties at low and high strains (1991)
Zhang, J., Andrus, R.D., Juang, C.H.: Normalized shear modulus and material damping ratio relationships. J. Geotech. Geoenviron. Eng. **131**(4), 453–464 (2005)

Effect of Width Variation of Liquefiable Sand Lens on Surface Settlement Due to Shallow Tunneling

Mohammad Azadi[1] and Lindsey Sebastian Bryson[2(✉)]

[1] Department of Civil Engineering, Qazvin Branch, Islamic Azad University, Qazvin, Iran
Azadi@qiau.ac.ir
[2] University of Kentucky, Lexington, KY, USA
sebastian.bryson@uky.edu

Abstract. There is a lot of research on the effects of earthquake loading on the forces and displacements of underground structures. Some researchers have investigated the influence of loose sand lenses that form due to geological processes in the saturation areas. However, an assessment of the influence of the sand lens dimensions on underground structures such as tunnels has rarely been assessed. Thus, the attempt of this paper is an assessment of the effects of width variation of liquefiable sand lens on surface settlement due to shallow tunneling. In this research, FLAC-3D software has been used to model the pore pressure changes and surface settlement during the liquefaction of the sand lens. According to the results, an increase in the sand lens width tends to have a direct correlation to the increase in surface settlement.

1 Introduction

Sometimes during earthquakes, apparent non-liquefiable soils show large and asymmetric deformations. This phenomenon occurs adjacent to rivers, seas and oceans. The geological investigations in these areas show that loose sand lenses are susceptible to liquefaction even though the surrounding soils have a good strength. These lenses may sustain the soil strength under static load, but under dynamic loading, they will lose their strength rapidly and change to semi-liquid. In this case, superstructures and also underground structures may be severely damaged. Tunnels constructed in this layer may be affected by the liquefaction as well. In this condition, the liquefaction may cause settlement in the ground, deformation in the tunnel shield, buoyancy in the underground buildings, reduction in bearing capacity and increases in lateral spread and pore pressures. Several studies have been carried out on the liquefaction of sand lenses under different conditions (Holchin and Vallejo 1995; Shokri 1996; Beheshti 1998; Vallejo 1998). Vallejo (1998) investigated the effects of sand lens in clay soil and calculated the failure angle of sand lens during a dynamic load. The results show that a tension crack of 70.5° is caused by the sand lens liquefaction in one side and the shear fractures are developed on the other side (Fig. 1).

D. Choudhury et al. (Eds.): GeoMEast 2018, SUCI, pp. 155–163, 2019.
https://doi.org/10.1007/978-3-030-01920-4_13

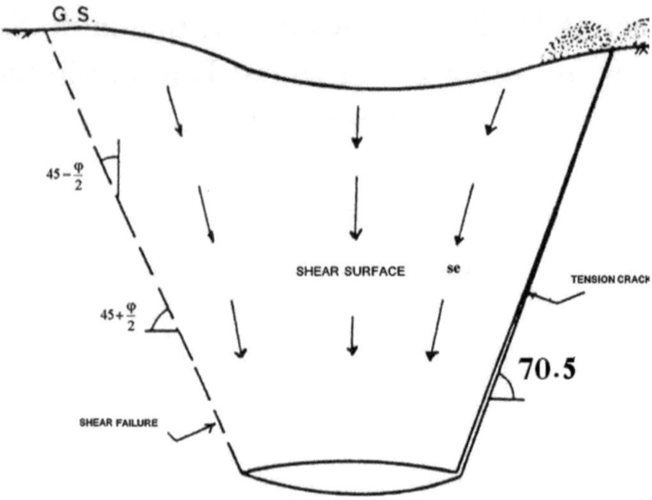

Fig. 1. Liquefaction effects of the sand lens on surrounding area (Vallejo 1998).

Other researchers have also investigated the ground deformation behavior associated with sand lens. Shokri (1996) used elastic theory to evaluate sand lens liquefaction and Beheshti (1998) used the NISA finite element software to evaluate the phenomenon. Beheshti (1998) concluded that an elastic- perfectly plastic model produced a good approximation of observed results. Thus, the Mohr-Coulomb model is assumed to be applicable to the current study.

In addition to the research of sand lens liquefaction, numerous studies have been conducted about the excavation of tunnels in the soil environment (Ding et al. 2011; Jones and Hunt 2011; Kuo et al. 2011; Shin et al. 2011; Lai et al. 2012; Unutmaz 2014; Zhang et al. 2015). But, there is little research on the effect of liquefiable soil on tunneling and underground structure; especially in the case of liquefaction of sand lens in the non-liquefiable soils. Therefore, the effects of liquefiable sand lens on tunneling in the urban areas are assessed in this paper. In this study, the effects of the liquefaction of a sand lens in different width on the tunnel lining have been investigated using the FLAC-3D software. The selected elements for the soil medium are cubic and for the tunnel lining are beam elements.

1.1 Development of the Reference Models

In this study, there are two reference models (mesh generation has been shown in Fig. 2). In Reference Model No. 1, a tunnel of 6.9 m diameter is excavated at a depth of 10 m inside a clay deposit (according to Fig. 3). The depth of bed rock is in the bottom of the model which is 20 (m). To assess the liquefaction conditions, it is necessary the soil to be saturated. Thus, the water table is leveled with the ground surface. The soil properties is shown in the Table 1.

Reference Model No. 2 is similar to the Reference Model No. 1, but the tunnel is surrounded by a cubical sand lens with a width of 10 m. Figure 4 and Table 1 show

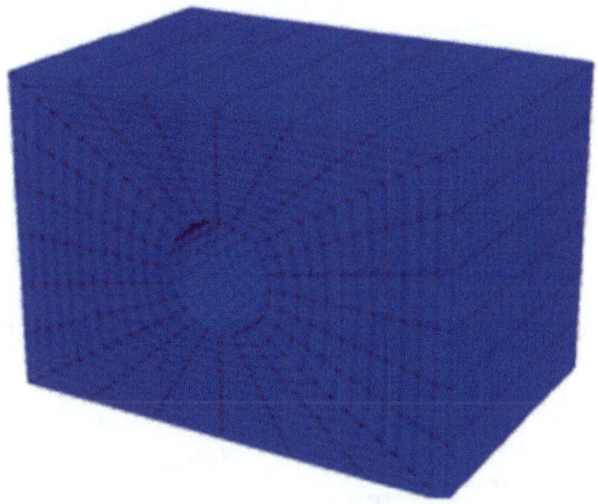

Fig. 2. Mesh generation in the reference models

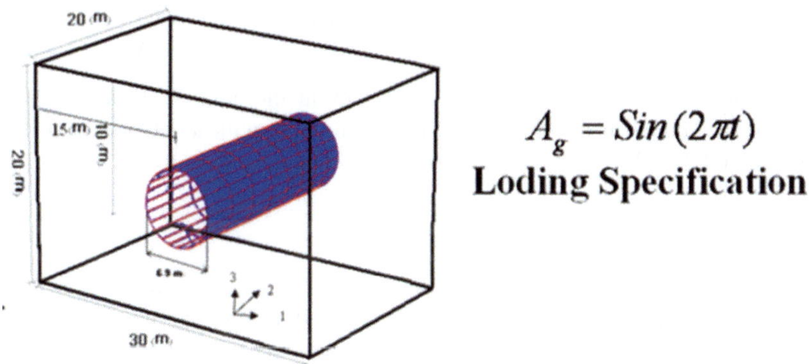

Fig. 3. Reference model No. 1 and loading specification in the model

Table 1. Soil properties

Soil property	Unit	Reference model No. 1	Reference model No. 2
G (Shear Modulus)	MPa	25	20
B (Bulk Modulus)	MPa	50	30
φ	Degree	30	25
C	kPa	0	0
γ_d	kN/m3	17	15
K	m/sec	10^{-6}	10^{-4}

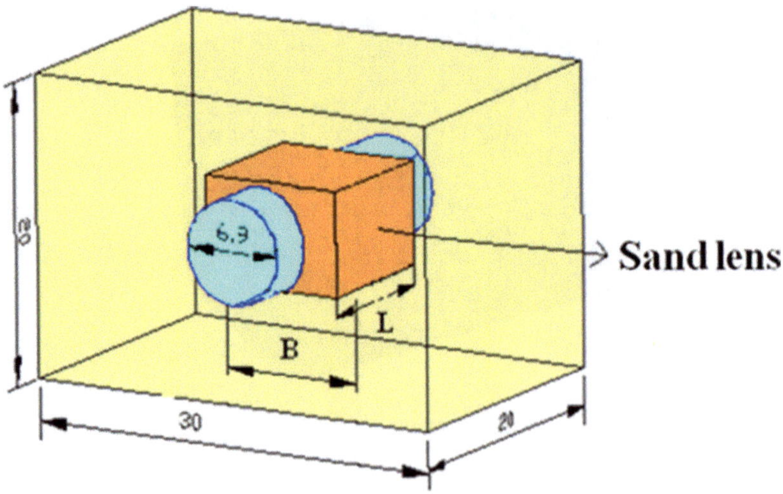

Fig. 4. Reference model No. 2 (including the sand lens)

this reference model and properties of sand lens in the area. A dynamic load was applied to the model in the form of a sinusoidal wave with amplitude of 0.1 g and frequency of 1 Hz. The loading function was defined according to Eq. 1 as:

$$\ddot{u}_g = \sin(2\pi t) \tag{1}$$

Where, $\ddot{u}_g$ is the acceleration from bedrock to the ground surface, t, is the duration of dynamic load. For the study, the duration of the dynamic load was assumed to be 10 s. This time is considered in such a way that the process of the analyses leads to a steady state and the impact of the dynamic load are seen completely.

In this study, the soil behavior was modeled by the Finn model (Byrne 1991). This model is suitable to assess the variations of the volumetric strain. This model is a modified version of the Mohr-Coulomb model and assumes the relationship between variations of volumetric strain increment, and cyclic shearing strain, γ can be described by Eq. 2, given as:

$$\Delta\varepsilon_{vd} = C_1(\gamma - C_2\varepsilon_{vd}) + \frac{C_3\varepsilon_{vd}^2}{\gamma + C_4\varepsilon_{vd}} \tag{2}$$

Where C1, C2, C3 and C4 are coefficients that can be obtained from cyclic triaxial tests and are constant in the equation (Martin et al. 1975). Based on studies performed by Pashangpishe (2004), values selected for the above coefficients are 0.79, 0.52, 0.2 and 0.5, respectively.

1.2 Comparison of the Results of the Reference Models

The liquefaction of a sand lens surrounding the tunnel may alter the pore pressures and effective stresses leading to new forces and deformations in soil and buried structure. Figures 5 and 6 show the increase of pore pressures in a point in the reference models. By comparing the pore pressure changes in these reference models, it becomes clear that the pore pressure due to liquefaction of the sand lens has increased in the Reference Model No. 2.

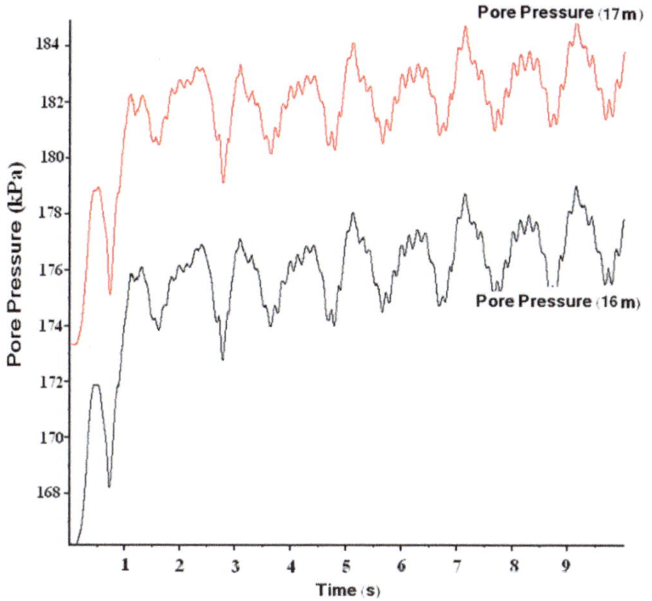

Fig. 5. Excess pore pressure in the reference model No. 1 (Under the tunnel axis).

Since the pore pressure increases as the number of load cycles increase, the forces acting upon the tunnel lining also may change while the cyclic loading continues. Also, since the pore pressure increases as the number of load cycles increase, the forces acting upon the tunnel lining also may change while the cyclic loading continues.

The changes of bending moment versus time are plotted in Fig. 7. It is quite evident that this parameter increases by increasing the time of dynamic loading. The results of the analyses show that the more loading time, the more axial forces and bending moments of tunnel lining are induced. According to the results, the maximum bending moment in the lining changes from 14.3 ton-m in the Reference Model No. 1 to 11.8 ton-m in the Reference Model No. 2. It may be attributed to the increase of soil displacements due to the liquefaction of the sand lens. The surface settlements show some 16% increase, as compared with Reference Model No. 1.

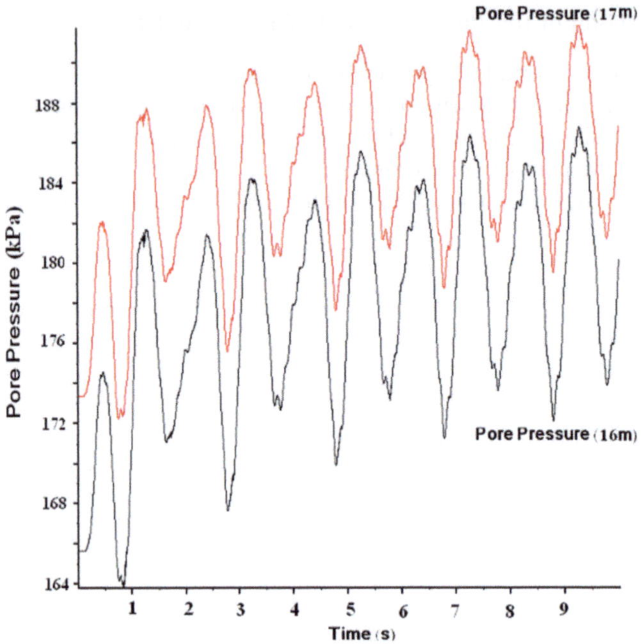

Fig. 6. Excess pore pressure in the reference model No. 2 (Under the tunnel axis).

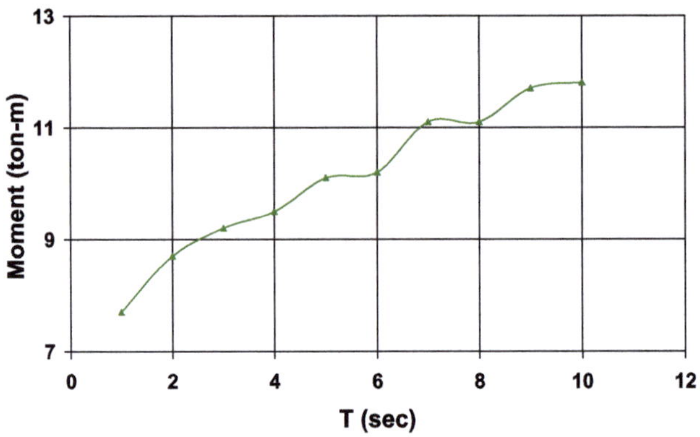

Fig. 7. The maximum bending moment in the tunnel lining versus cyclic loading time

1.3 Assessment of the Sand Lens Width

In order to assess the effects of the sand lens width, a sand lens with a width of 10 m is regarded in accordance with the properties of the Reference Model No. 2. Since the lens center coincides with the tunnel axis in this reference model, three other models are taken in to account by changing the sand lens width, relative to its initial position. Then the liquefaction effect of the sand lens is assessed under cyclic loading and the deformation and moments induced in the tunnel lining are evaluated. Figure 8 and Table 2 show the surface displacement and maximum bending moments in the different cases.

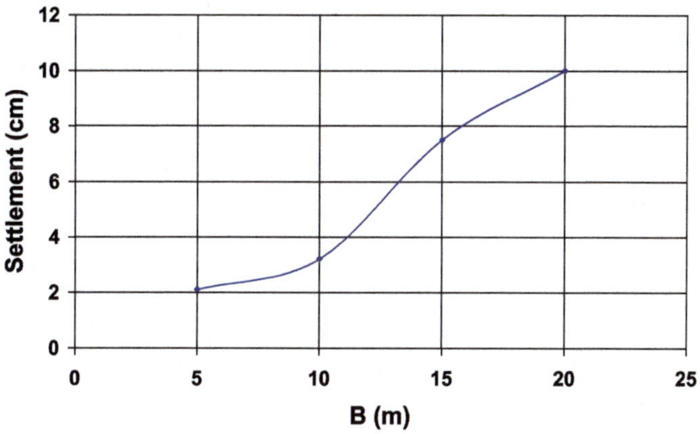

Fig. 8. Variations of the surface settlement in location of tunnel axis for different width of sand lens length

Table 2. Bending moment and surface settlement for different width of sand lens.

Width of the sand lens (m)	Surface settlement below the tunnel axis (cm)	Maximum bending moment in the tunnel lining (ton-m)	Percent of the surface settlement to the surface settlement of the reference model No. 2	Percent of the maximum bending moment to the maximum bending moment of the reference model No. 2
5	2.1	9.5	−34.3	−7.7
10 (Ref model No. 2)	3.2	10.3	0	0
15	7.5	10.9	134.3	5.8
20	10	16	212.5	55.3

Regarding the table, it can be seen that there is a definite link between the sand lens width and maximum bending moment. This link is remarkably after the width 15 m. According to the results, percent of the maximum bending moment in the width 15 m to the maximum bending moment of the Reference Model No. 2 (width 10 m) is almost 5.8%. However, this amount for the width 20 m is about 55%. Thus, it should be noted that the effect of increasing the sand lens width by 15 (m) on moments is negligible, but after this width, this effect can be sensible.

The results also show that increasing the lens width leads to the reduction in the surface displacements. The reason is the decrease in the liquefaction potential of the lens due to increasing the width. Referring to Table 2, when the width of sand lens is about 5 m, the surface settlement below the tunnel axis reaches to 2.1 cm, while the settlement increases sharply in case of having a wider than liquefiable sand lens at the same case. For the case when the width of sand lens is about 20 m, the ground settlement becomes 10 cm and it is about 3 times more than the settlement in the Reference Model No. 2.

2 Conclusions

In this paper, the effects of the liquefaction of a sand lens of different width on the tunnel lining have been investigated using the FLAC-3D software. Two reference models were defined and considered to represent the case in which the tunnel is located at a depth of 10 m from the ground surface. Then the effects of the surrounding area on the tunnel have been focused. At this stage, the width of the sand lens was altered and the influences of liquefaction of such lens at different cases relative to the tunnel on the surface settlements were investigated. Based on the investigations, during dynamic loadings, the pore pressure in the area is increased and the effective stress is decreased. The reduction in the effective stress causes an increase in horizontal and vertical displacements of the soil medium and results in extensive liquefaction.

The increase in the sand lens width causes the surface displacements to decrease. When the sand lens does not exist in the area, tunnel settlement reaches the minimum value. For the case when the width lens is about 20 m, the ground settlement becomes 10 cm and it is about 3 times more than the settlement in the Reference Model No. 2.

The other point of this study is the bending moment in the tunnel lining. According to the results, the effect of increasing the sand lens width by 15 m on moments is negligible, but after this width, this effect can be sizable.

References

Beheshti, Kh.: The investigation of the behavior of saturated sand lenses within the soil deposits under dynamic loading. Msc thesis, Submitted in Civil Eng. Faculty of Amirkabir University of Technology, Iran (1998)

Byrne, P.: A cyclic shear-volume coupling and pore pressure model for sand. In: Proceedings Second International Conference on Recent Advances in Geotechnical Earthquake Engineering and Soil Dynamics, St. Louis, Missouri (1991)

Ding, L., Ma, L., Lou, H., Yu, M., Wu, X.: Wavelet analysis for tunneling-induced ground settlement based on a stochastic model. Tunn. Undergr. Space Technol. **26**(5), 619–628 (2011)

Holchin, J.D., Vallejo, L.E.: The liquefaction of sand lenses due to cyclic loading. In: Proceedings 3rd International Conference on Recent Advances in Geotechnical Earthquake Engineering and Soil Dynamics, Vol. I, St. Louis, Missouri, p. 253 (1995)

Jones, S., Hunt, H.: Voids at the tunnel-soil interface for calculation of ground vibration from underground railways. J. Sound Vib. **330**(2), 245–270 (2011)

Kuo, K.A., Hunt, H.E.M., Hussein, M.F.M.: The effect of a twin tunnel on the propagation of ground-borne vibration from an underground railway. J. Sound Vib. **330**(25), 6203–6222 (2011)

Lai, J., Fan, H., Dong, Z.: Excavation effect of tunnel slope reinforced with anti sliding pile in cold regions based on 3D-FEM, Trans Teach Publication, Switzerland. Adv. Mater. Res. **413**, 170–173 (2012)

Martin, G.R., Finn, W.D.L., Seed, H.B.: Fundamentals of liquefaction under cyclic loading. J. Geotech. Div. **101**(5), 423–438 (1975)

Pashangpishe, Y.: Mechanism of soil deformation due to double lenses liquefaction and critical depth determination. Msc thesis, Submitted in Civil Eng. Faculty of Amirkabir University of Technology, Iran (2004)

Shin, Y.J., Song, K.L., Lee, I.M., Cho, G.Ch.: Interaction between tunnel supports and ground convergence-consideration of seepage forces. Int. J. Rock Mech. Min. Sci. **48**(48), 394–405 (2011)

Shokri: Evaluation of the liquefaction potential of sand lenses. Msc thesis, Submitted in Civil Eng. Faculty of Amirkabir University of Technology, Iran (1996)

Unutmaz, B.: 3D liquefaction assessment of soils surrounding circular tunnel. Tunn. Undergr. Space Technol. **40**, 85–94 (2014)

Vallejo, L.E.: The liquefaction of sand lenses during an earthquake. In: VonThun, J.L. (ed.) Proceedings of the Earthquake Engineering and Soil Dynamics II – Recent Advances in Ground Motion Evaluation, No. 20, ASCE, p. 493. ASCE Geotechnical Special Publication, New York (1998)

Zhang, Y., Yang, J., Yang, F.: Field investigation and numerical analysis of landslide induced by tunneling. Eng. Fail. Anal. **47**(Part A), 25–33 (2015)

Liquefaction Potential Analysis and Numerical Modeling - Container Terminal of Algiers Port

Mohamed Chikhaoui(✉), Mohamed A. Mehenni, Lynda Djerbal, and Walid Boulifa

LEEGO Laboratory, Faculty of the Civil Engineering, University of Sciences and Technology Houari Boumediene (USTHB), Ezzouar, Algiers, Algeria
mch_gcgl6@yahoo.ca

Abstract. The aim of the present paper is to study the liquefaction potential of the container terminal of Algiers port site and perform a numerical modeling of its dynamic response under Boumerdes earthquake. The soil liquefaction potential was analyzed using two methods. The first is an analytical method requiring a series of observations and empirical correlations carried out from in situ investigations, namely the SPT (Standard Penetration Test). In addition, the liquefaction safety factor obtained was verified according to the recommendations of the Eurocode rules and the Algerian earthquake regulation. The MSF coefficient was determined to obtain a corrected CSR value corresponding to a seismic motion with magnitude of 7 on the Richter scale. The second analysis is carried out using PLAXIS 2D (version 2016) software using soil parameters determined from standard triaxial and œdometric tests. The numerical modeling was established in order to study dynamic response and verify the liquefaction potential of Algiers port under Boumerdes earthquake taking into account the presence of the container static load.

1 Introduction

Following many earthquakes, a liquefaction of saturated granular soils has been observed. Indeed, the appearance of the liquefaction hazard in the soils can have devastating consequences on all the infra or the superstructure. We cite as examples:

- Impoundments reservoir collapses during the Niigata earthquake in Japan (1964),
- Liquefaction of soil during the El-Asnam earthquake, Algeria, 1980,
- Sand volcanoes during the superstition earthquake Hills, California (1987),
- Landslide of Monte Zandila in Valtelline (Lombardy, Italy), 1987,
- And recently, the Tohoku earthquake in Japan (2011), etc.

The north of Algeria, as several countries over the world, has experienced several extensive earthquakes that caused important buildings damages and people injuries. The Boumerdes of 21st May 2003 (Mw = 6.8) was the second most destructive land motion in northern Algeria during the last 50 years, after the El Asnam earthquake of 1980 (Ms = 7.3). The container terminal of Algiers port, situated at approximately 50 km from the Zemmouri earthquake epicentral location, was strongly damaged during this event. Given the economic and strategic importance of this port, located at

D. Choudhury et al. (Eds.): GeoMEast 2018, SUCI, pp. 164–173, 2019.
https://doi.org/10.1007/978-3-030-01920-4_14

the capital of Algeria, a good characterization of damages causes is necessary to select an appropriateness rehabilitation technique. In fact, this site is characterized by several geotechnical constraints, which are, principally, low bearing capacity, high compressibility and liquefaction susceptibility of the saturated loose sand layers of 12 m thick (Boulifa et al. 2017).

For assessing the liquefaction potential of soils, several methods have been proposed (Seed and Idriss 1982, 1970, 1968). Over time, the simplified method established by (Seed and Idriss 1982, 1970, 1968) has been modified several times and has become the most commonly used in this field. The method of calculating the liquefaction potential originally proposed by (Seed and Idriss 1982, 1970, 1968) was updated by (Youd et al. 2001), and more recently by (Boulanger and Idriss 2006), but the essence of the procedure has remained broadly unchanged since the original publication. Moreover, the liquefaction potential is a function of the ratio of the mean cyclic shear stress acting on horizontal planes of the ground during the seismic load and that of the effective stress acting on the soil before the load (Youd et al. 2001).

Considering the widespread use of the method of (Youd et al. 2001), it had been used to evaluate the liquefaction potential of Algiers port site (container terminal zone) exploiting SPT test results. In order to give a batter analysis of the liquefaction phenomena in this site, PLAXIS 2D (version 2016) software has been used. The numerical modeling of the site takes into account the soil parameters determined from standard triaxial and œdometric tests (Boulifa et al. 2017). This numerical analysis takes also into account the presence of a uniform static load due to the container weight.

1.1 Site Presentation

Characterized by a storage capacity of 282,000 m^2, the Algiers port is located in the western zone of the bay of Algiers and covers an area of 126 ha. The geological map of Algiers department shows that a large area of the coastal zone of this site is covered by Quaternary terrains mainly composed of sand, gravel and sandy clays materials. For that, a series of geotechnical investigations were carried out at the site of the container terminal following the significant damages that the latter underwent during the Boumerdes earthquake. The companion consisted of 27 SPT tests, including 15 with core drilling subjected to laboratory tests, 13 CPT tests and eight pressiometric tests (Fig. 1). The results of the laboratory tests are summarized in Table 1.

According to the triaxial and œdometric tests, the sand layer of Algiers port has the behavior of loose sand overconsolidated and perfectly dilatant criterion under the seismic effect.

1.2 Liquefaction Potential Analysis

In order to estimate the liquefaction potential at the container terminal site of the Algiers port, the results of the standard dynamic penetration tests (SPT) were used in this study, using the method of (Youd et al. 2001).

In addition, this method (Youd et al. 2001) is really based on the methodology proposed by (Iwasaki et al. 1982). It permit the evaluation of the liquefaction potential

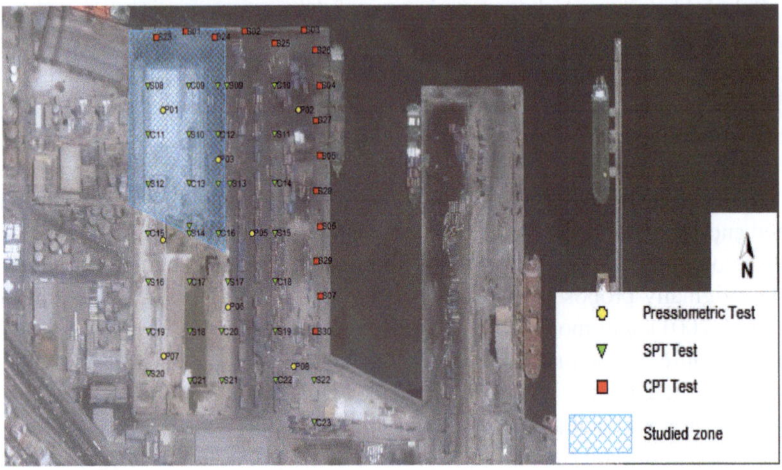

Fig. 1. In situ tests implantation

of soils by calculating the safety factor (Fs), which is considered as the ratio of the cyclic resistance ratio (CRR) to the cyclic stresses ratio (CSR) according to the Eq. 1.

$$FS_{liquefaction} = \frac{CRR}{CSR} < 1.25 \tag{1}$$

The CRR is calculated using the following Eq. (2):

$$CRR_{7.5} = \frac{1}{34 - (N_1)_{60}} + \frac{(N_1)_{60}}{135} + \frac{50}{\left[10 \times (N_1)_{60} + 45\right]^2} - \frac{1}{200} \tag{2}$$

$(N_1)_{60}$: is the corrected penetration resistance. This parameter is calculated using the formulation proposed by (Youd et al. 2001):

$$(N_1)_{60} = N_{30} \times \frac{100}{60} \times C_N \times C_B \times C_S \times C_E \times C_R \tag{3}$$

C_N: the standardized relatively to the overburden pressure factor calculated according to the equation proposed by (Liao and Whitman 1986):
$Cn = (Pa/\sigma'_v)^{0.5}$, with Pa (Atmospheric pressure) and the σ'_v (effective vertical stress)
C_E: the correction for hammer energy ratio
C_B: the correction for borehole diameter
C_R: the correction factor for rod length
C_S: the correction for samplers with or without liners

Table 1. Soil layers parameters according to the laboratory tests

Depths (m)	0–0.24	0.24–0.44	0.44–0.79	0.79–1.59	1.59–17.43		17.43–30
Materials	Bituminous concrete (BB)	Bituminous gravel (GB)	Untreated gravel (GNT)	Tuff	El-Hadjar Sand		Marly clay
Behavior laws	Linear elastic	Linear elastic	Mohr Coulomb	Mohr Coulomb	Hardening Soil Small		Soft Soil
Drainage mode	No porous	No porous	Undrained	Undrained	Undrained		Drained
γ_d (KN/m^3)	23.5	23.5	20	20	17.66		15.2
γ_{sat} (KN/m^3)	–	–	22.5	22	19.62		19.32
C'(kPa)	–	–	0.1	1	1		35
$\varphi'(\circ)$	–	–	30	29.5	32		27
E'(KPa)	4.32^{E6}	7.5^{E6}	4.8^{E5}	3.8^{E4}	9.163^{E3}		$2{,}60^{E+04}$
ν	0.35	0.35	0.35	0.35	0.30		0.30
e_0	–	–	–	–	0.466		0.661
C_c	–	–	–	–	0.046		0.08
C_s	–	–	–	–	4.29^{E-3}		0.02
K (m/j)	–	–	8.64^{E-4}	8.64^{E-4}	X = 0.864	Y = 0.0864	1.503^{E-5}

Reference (Youd et al. 2001) proposed a correction of $(N_1)_{60}$ in order to obtain an equivalent clean sand value $(N_1)_{60CS}$ given by Eq. (4):

$$(N_1)_{60CS} = \alpha + \beta \times (N_1)_{60} \tag{4}$$

With:

$$\begin{aligned}
&\alpha = 0 \quad for \quad FC \leq 5\% \\
&\alpha = e^{\left[1.76 - \left(190/FC^2\right)\right]} \quad for \quad 5\% < FC < 35\% \\
&\alpha = 5 \quad for \quad FC \geq 35\% \\
&\beta = 1 \quad for \quad FC \leq 5\% \\
&\beta = \left[0.99 + \left(\frac{FC^2}{1000}\right)\right] \quad for \quad 5\% < FC < 35\% \\
&\beta = 1.2 \quad for \quad FC \geq 35\%
\end{aligned}$$

FC: represent the fine content

The CSR defines the seismic demand and is expressed as Eq. (5):

$$\mathrm{CSR} = \frac{\tau_{av}}{\sigma_{v0'}} = 0.65 \times \frac{a_{max}}{g} \times \frac{\sigma_{v0}}{\sigma_{v0'}} \times r_d \tag{5}$$

a_{max}: Peak horizontal ground acceleration (given in (RPA 99 2003) in function of the seismic zone and building importance. For this case: $a_{max} = 0.30 \times g$), g: acceleration due to the gravity, σ_{v0}: vertical total stress at the simple depth and $\sigma_{v0'}$ vertical effective stress at the simple depth.
r_d: Stress reduction factor estimated using Priebe (1998) equation in function of the simple depth (6):

$$r_d = 1 - 0.015 \times Z \tag{6}$$

Z: simple depth.

Then, the CSR values have been divided by the magnitude scaling factor (MSF), which is calculated using the formulation proposed by (Youd et al. 2001) as follows:

$$MSF = \left(\frac{M_W}{7.5}\right)^{-2.56} \tag{7}$$

The liquefaction potential of the container terminal of Algiers port was evaluated for an earthquake of 7.0 magnitude, to consider a safety margin with respect to 21st

May 2003 Boumerdes (Mw = 6.8) earthquake, which is the strongest earthquake recorded in Algiers region. The obtained results allowed confirming the liquefaction potential of this site. Otherwise, the results obtained for the different SPT test have been summarized in Table 2.

Table 2. Example of the potential liquefaction estimated using the SPT test

Depth (m)	$CRR_{7,5}$	CRR_7	rd	CSR	F_7	$Liquefaction_7$
2.00	0.1208	0.1570	0.9700	0.1887536	0.8317908	L
2.50	0.1711	0.2225	0.9600	0.208769	1.0655717	L
3.00	0.1487	0.1933	0.9600	0.2265224	0.8533098	L
3.50	0.1471	0.1912	0.9500	0.2386594	0.8012853	L
4.00	0.1013	0.1317	0.9400	0.2481848	0.5306489	L
4.50	0.1066	0.1386	0.9300	0.2556815	0.5420592	L
5.00	0.0900	0.1170	0.9300	0.2644144	0.4424962	L
5.50	0.1352	0.1757	0.9200	0.269091	0.653029	L
6.00	0.1299	0.1689	0.9100	0.2726992	0.6192558	L
6.50	0.1242	0.1615	0.9000	0.2754228	0.5862627	L
7.00	0.1078	0.1401	0.9000	0.2805225	0.499533	L
8.00	0.0978	0.1272	0.8800	0.2827977	0.4497747	L
8.14	0.1365	0.1774	0.8800	0.2838579	0.6249956	L
8.50	0.1442	0.1874	0.8700	0.2832017	0.6618821	L
9.80	0.1515	0.1969	0.8500	0.2845046	0.6921142	L
10.00	0.1153	0.1499	0.8500	0.2855591	0.5250396	L
10.30	0.1058	0.1375	0.8500	0.2870776	0.4791077	L
10.50	0.1173	0.1525	0.8400	0.2846613	0.5357284	L
10.80	0.1149	0.1494	0.8400	0.2860477	0.5222008	L
11.30	NL	NL	0.8300	0.2847903	–	NL
11.36	NL	NL	0.8300	0.2850374	–	NL

1.3 Numerical Modeling

The numerical modeling of the container terminal of Algiers port is conducted using PLAXIS 2D software. It takes into account the static loading of the container and seismic loading.

Model Presentation

The modeled profile is presented in the Fig. 2. It contains two layers of in situ soils and four layers constituting the pavement. The parameters considered for these soils are given in Table 1. The soil profile is defined from the core drilling C_{09} example and extends over a length of 240 m and 30 m elevation.

Displacement and dynamic boundaries conditions are needed to represent the far-field behavior. Appropriate displacement boundary conditions allow to represent body displacements in the limits of the model. These boundaries have been defined as: the vertical geometric lines for which the abscissa (x) is equal to the smallest or the largest

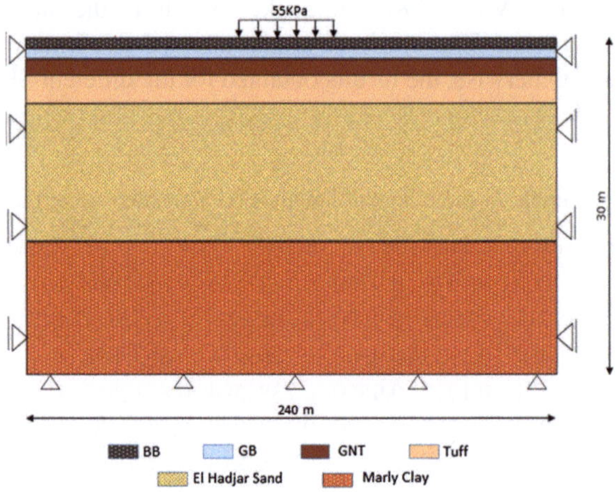

Fig. 2. Soil profile of container terminal of Algiers port

of the abscissa of the profile (x = 0 m and x = 240 m) are blocked in the horizontal direction (Ux = 0) and the geometric lines for which the distance (y) is equal to the smallest ordinate of the model is fully blocked (Ux = Uy = 0).

Dynamic boundaries can simulate the far-field behavior during dynamic excitation by absorbing stresses increment and avoiding inappropriate wave reflections. The dynamic boundary conditions chosen are as: the vertical geometric lines for which the abscissa is equal to the smallest or the largest abscissa of the profile are free field one to absorb the outgoing wave energy, the surface of the model is free and the base of the model is compliant one. The compliant base boundary has been adopted to simulate the continuation of waves into the deep soil with minimum reflection at the boundary. The mesh chosen for this calculation is constituted by 15,702 triangular elements with 15 nodes (Fig. 2).

The Dynamic Calculation Methodology

The numerical modeling of Algiers port liquefaction under Boumerdes (Algeria) earthquake is performed taking into account the presence of a static load due to the container weight, which is considered as a uniform load (Fig. 3).

The methodology explained in Fig. 3, allows to observe the main steps conducted to perform this numerical analysis.

The dynamical analysis of Algiers port site, effectuated using PLAXIS 2D software under 21st May 2003 Boumerdes earthquake conditions, is conducted taking into account the accelerogram of Hussein Dey station, which is nearest to the studied site (Table 3).

The component EW of this station has been chosen because it is the highest one. It is characterized by a peak ground velocity PGV = 0.16 m/s and Peak ground acceleration PGA = 0.27 g and taken 22 s (Fig. 4) (Laouami et al. 2006). The earthquake acceleration has been applied at the base of the model in the horizontal direction.

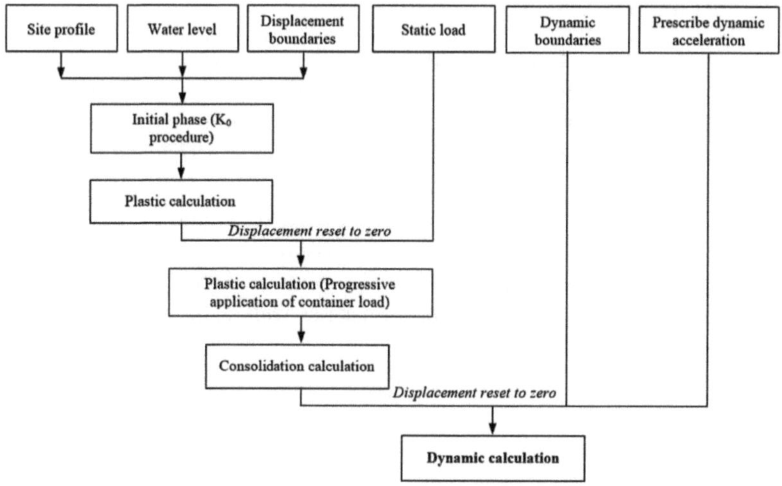

Fig. 3. Numerical modeling organogram for the dynamic calculation

Table 3. Recorded peak ground acceleration velocities and displacement of Boumerdes earthquake at Hussein Dey station (Latitude 36.74° – Longitude 3.096°)

Direction	PGA (g)	PGV (cm/s)	PGD (cm)
East–West	0.27	16.5	3.9
Vertical	0.09	7.7	1.8
North–South	0.23	9.1	2.7

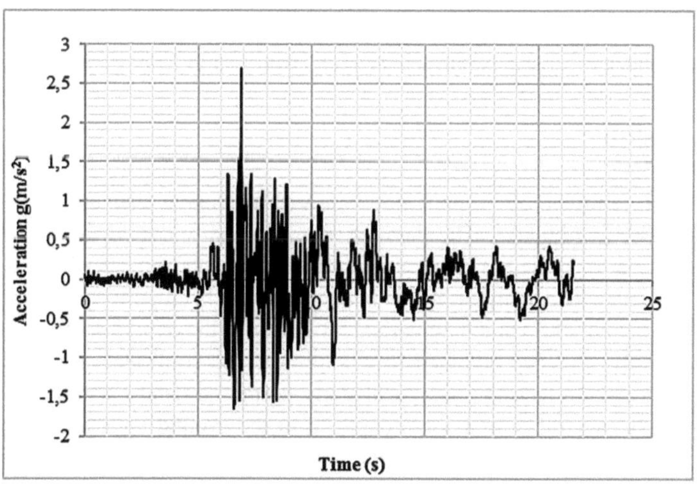

Fig. 4. EW Accelerogram component of Hussein Dey station during 21st May 2003 Boumerdes earthquake (Laouami et al. 2006)

1.4 Results and Discussions

Liquefaction of granular materials is represented by a progressive increase of the pore pressure until a partial or total loss of the soil shear resistance. This phenomena is then expressed by an important decrease or weakening of the effective stress according to the ascertainment of (Casagrande 1975; Castro and Poulos 1977) and (Hareb 2009).

Figure 5, representing soil stress and pore pressure calculated at 10 m depth in the El-Hadjar sand layer, allows to observe that the total stress curves (in three principal directions) are intersected by the excess pore pressure curve. So, this result shows an annulment of the effective stress of El-Hadjar sand and confirms the liquefaction observed during the 2003 Boumerdes earthquake as well as the results obtained using the method of (Youd et al. 2001).

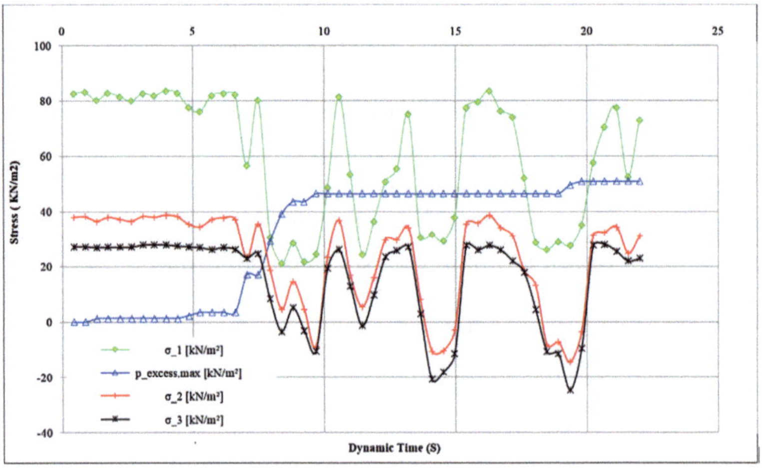

Fig. 5. Total stresses and excess pore pressure expressed in term of the dynamic time

2 Conclusion

The container terminal of Algiers port suffered extensive damages following the Boumerdes 2003 earthquake. The damage was principally due to the loose saturated sand liquefaction expressed at the ground surface by displacements of about 30 cm to 40 cm in this zone. The analysis of this phenomena has been performed using analytic method (Youd et al. 2001) and a numerical modeling of the dynamic soil response with PLAXIS 2D software. The obtained results allowed the confirmation of the sand liquefaction potential. In fact, the analytical study permitted determination of liquefiable layer ranging between 2 m and 11 m depth from the ground level and the numerical study led to show an annulment of the effective stresses of the loose sand in the three principal directions. Besides, these results confirm the liquefiable character of sand layer and let to explain the important disorder caused to Algiers port during Boumerdes earthquake. It constitutes a first characterization study, which can offer good bases for the future reinforcement analyses of this site.

References

Boulanger, R.W., Idriss, I.M.: Liquefaction susceptibility criteria for silts and clays. J. Geotech. Geoenviron. Eng. **132**(11), 1413–1426 (2006). 10.1061/(ASCE)1090-0241(2006)132:11(1413)

Boulifa, W., Mehenni, M.A., Chikhaoui, M.: Amélioration de sol du port d'Alger par colonnes ballastées. Travaux publics. Géotechnique, Faculté de génie civil. Université des Sciences et de la Technologie Houari Boumediene USTHB. Usthb.dz, Bab Ezzouar, Alger, Algérie (2017)

Casagrande, A.: Liquefaction and cyclic deformation of sand: a critical review. In: 5th Panamerican Conference on Soil Mechanics and Foundation Engineering édition (1975)

Castro, G., Poulos, S.: Factors affecting liquefaction and cyclic mobility. Journal of Geotechnical Engineering édition (1977)

Hareb, H.: Influence de l'histoire des prechargements draines sur le comportement a la liquefaction du sable d'hostun tres lache. École doctorale des sciences pour l'ingénieur de lyon, Laboratoire GéoMatériaux École Nationale des Travaux Publics de l'Etat (ENTPE), Université de Béjaïa, Algérie (2009)

Iwasaki, T., Tokida, K., Tatsuoka, F., Watanabe, S., Yasuda, S., Sato, H.: Microzonation for soil liquefaction potential using simplified methods. In: 3rd International Conference on Microzonation, Seattle, vol. 3, pp. 1310–1330 (1982)

Laouami, N., Slimani, A., Bouhadad, Y., Chatelain, J.-L., Nour, A.: Evidence for fault-related directionality and localized site effects from strong motion recordings of the 2003 Boumerdes (Algeria) earthquake: consequences on damage distribution and the algerian seismic code. Soil Dyn. Earthquake Eng. **26**(11), 991–1003 (2006). https://doi.org/10.1016/j.soildyn.2006.03.006

Liao, S.S.C., Whitman, R.V.: Overburden correction factors for SPT in sand. J. Geotech. Eng. **112**(3), 373–377 (1986). https://doi.org/10.1061/(ASCE)0733-9410(1986)112:3(373)

RPA 99: Règlement Parasismique Algérien. DTR-B.C.2.48, Ministry of Housing and Urban Planning, Algeria. C.G.S. Centre de Géophysique et de Sismologie (2003)

Seed, H.B., Idriss, I.M.: Landslides during earthquakes due to liquefaction. J. Soil Mech. Found. Div Earthquakes Liquefaction Sand Soils **94**(5), 1055–1122 (1968)

Seed, H.B., Idriss, I.M.: Simplified procedure for evaluating soil liquefaction potential. J. Soil Mech. Found. Div Earthquakes Liquefaction Sand Soils **97**(SM9, PROC PAPER 8371), 1249–1273 (1970)

Seed, H.B., Idriss, I.M.: Ground Motions and Soil Liquefaction during Earthquakes. Engineering Monographs on Earthquake Criteria, Structural Design, and Strong Motion Records, vol. 5, Earthquake Engineering Research Institute, Berkeley (1982)

Youd, T.L., Idriss, I.M., Andrus, R.D., Arango, I., Castro, G., Christian, J.T., Dobry, R., et al.: Liquefaction resistance of soils: summary report from the 1996 NCEER and 1998 NCEER/NSF workshops on evaluation of liquefaction resistance of soils. J. Geotech. Geoenviron. Eng. **127**(10), 817–833 (2001). https://doi.org/10.1061/(ASCE)1090-0241(2001)127:10(817)

Research on Impact Vibration Test for Port Structure in Viet Nam

Nguyen Thi Bach Duong(✉), Nguyen Huu Hung, and Nguyen Anh Dan

Urban Transport and Marine-Coastal Engineering Section, Faculty of Civil Engineering, University of Transport and Communications, Hanoi, Vietnam
{ntbachduong, nhhunggttp, nadan}@utc.edu.vn

Abstract. IVT (Impact Vibration Test) was used to determine the natural vibration frequency of the structure in order to evaluate the structural health. This is a non-destructive testing method that is simple to carry out, accurate measurement of natural vibration frequencies, low cost and short time. This article introduces, studies and applies this method for health assessment of port structures in Viet Nam.

1 Introduction

In general, the load test, structure evaluation, health assessment of engineered structures can apply methods: static load test or/and dynamic load test. However, those works are very difficult, even they are impossible to be done for port structures. In particular, evaluation of horizontal load-bearing capacity is made by ship berthing and ship motion. In many cases the cost could be very expensive, not really effective.

In some developed countries, non-destructive load testing (shock, vibration and impact testing) are used increasingly to assess the health of sub-structure and pile foundation. Impact vibration test (IVT) has been applied to examine for 300 structures in Japan. In Vietnam, Standard 22TCN 243-98 also sets criteria for the sub-structure by dynamic method. However, evaluation criteria in 22TCN 243-98 are not clear. In fact, the many cases are not enough bearing strength or load-bearing capacity but the evaluations of 22TCN 243-98 are still satisfy.

The paper introduces, research the application of IVT method such as: making of impact load, designing of theoretical dynamical analysis model and determining of nature frequency by professional experience method as well as the application of these studies to the health assessment of the port structure in Viet Nam.

2 Research on Impact Vibration Method for Port Structure

2.1 Impact Vibration Test

This method evaluates the natural oscillation frequency which is received by structural impact and forced to oscillate. Comparing the results of its natural vibration frequency received with available data, we can assessment the strong of structure by using natural

D. Choudhury et al. (Eds.): GeoMEast 2018, SUCI, pp. 174–182, 2019.
https://doi.org/10.1007/978-3-030-01920-4_15

oscillations analysis from oscillations measured results. We can determine the structural health status. This method has some advantages as below:

- There is a non-destructive test which is based on evaluation for natural-vibration frequency.
- That is specified from structure behavior when load weight is struck to structure (impact load).
- The test is performed very simple, vibration frequency is exactly specified.
- Evaluation gets even foundation structure where you can't see it with the naked eye.
- Evaluation gets quantitative with high exact.

2.2 Procedure of Impact Vibration Test

The contents of the diagnostic methods of substructure by IVT include the following steps:

Step 1: Data collection from pulse test.
Step 2: Vibration data processing of structure (Furier Fast and Huang Hilbert Transform, refer Fig. 1)

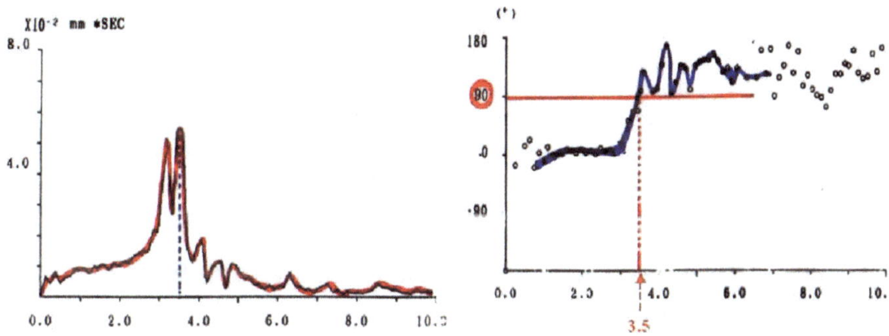

Fig. 1. Processing results with Fourier Spectrum

Step 3: Specify health index of structure κ

$$\textbf{Index of Soundness}\ (\kappa) = \frac{\textbf{Measured Natural Frequency}}{\textbf{Standared Value of Natural Frequency}}$$

Step 4: Compare and assess health index of structures κ with evaluation criteria and conclude health situation of inspected structure (show in Table 1, refer Japan standards)

Table 1. Compare results evaluation based on Japan standard

Index of soundness κ	Level	Measure
$\kappa \leq 0.70$	A1	May cause a danger if there is great external load applied, need to implement other investigation, need to reinforce
$0.70 < \kappa \leq 0.85$	A2	Need to continue pay attention to the decrement in the natural frequency
$0.85 < \kappa \leq 1.00$	B	There is almost no problem at present
$1.00 < \kappa$	S	The structure is completely healthy at present

2.3 Soundness Evaluation of Sub-structure

To diagnostic, evaluate exploitation technical situation of structure, we can use one of the methods as below:

The first method: It's comparative with natural-vibration frequency which is measured in the past.

The second method: It's comparative with standard value of natural-vibration frequency (this standard value may be specified follow formulas in Japan standard).

The third method: It's comparative with design value of natural-vibration frequency. Otherwise, from natural-vibration frequency determined via measurement results, evaluate hardness of pier and spring factor of soil, comparative these values with design values.

In case of the segments of port structure must to apply 2rd method in Viet Nam. The standard value of natural-vibration frequency is specified follow formulas in Japan standard.

3 The Health Assessment of Port Structure Evaluation by Impact Vibration Method

3.1 Making Impact Load

Plan 1: Can choose one of two ways (refer Fig. 2).

- Installation system support load weight by steel frame, depending on the dimensions of specific surface of transverse frame to design suitable frame.
- Install the weight hanger using mobile crane, ensuring the reach of the crane is long enough. The height of hanger is 2 m higher than the floor elevation.

Plan 2: Using mobile crane (refer Fig. 3).
Plan 3: Using the tugboat (refer Fig. 4).
Plan 4: Using a vessel to establish anchoring loads (refer Fig. 5).
Plan 5: Establish impact forces using movable gantry crane (refer Fig. 6).

Fig. 2. Establish lateral impact forces using the weight

Fig. 3. Establish lateral impact forces using mobile crane

Fig. 4. Making of an impact force by tugboat

Fig. 5. Using a vessel to establish anchoring loads

Fig. 6. Establish impact forces using movable gantry crane

3.2 Theoretical Dynamic Analysis Model

The theoretical dynamic analysis model of port segments is shown in Fig. 6, it is made by eigen value analysis in SAP 2000. This is a frame structure consisting of transverse and longitudinal beams and shell elements. Boundary condition is a type of elastic connection between the pile and the ground as shown by springs. The mass of the crane and the infrastructure is assigned on the top of the structural system. For segment without crane this crane load will not be calculated in the numerical model (refer Fig. 7).

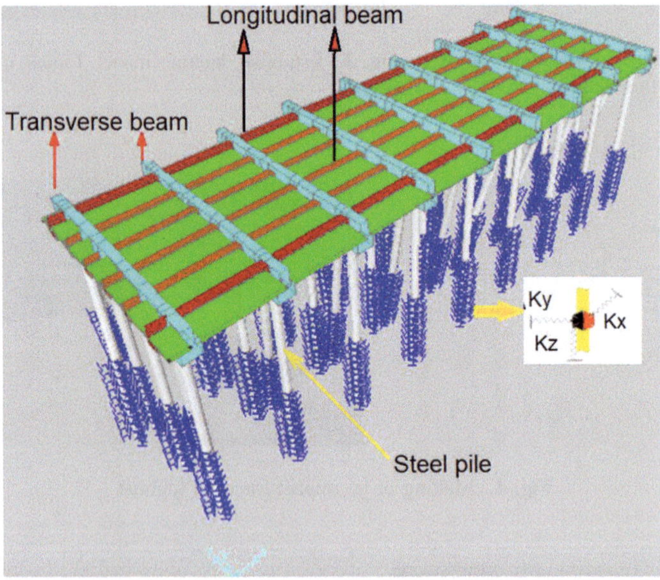

Fig. 7. Dynamic analysis model of port structure

Horizontal and vertical spring constants K_x, K_y, K_z are estimated using following formulas:

$$K_{xi} = Dxl_ixk_{xi}; K_y = Dxl_ixk_{yi}; K_{zi} = \pi Dxl_ixk_{zi}$$

When pile is compressed, besides 3 springs at the middle of pile, there is a spring at the pile tip with a spring constant K_{mz}

$$K_{mz} = A_m \times k_{mz}$$

In which:
Determination of reaction coefficients according to the OCDI 2009 standard

Horizontal reaction coefficients are estimated according to methods of Chang and Port and harbour research institute (PHRI):

$$k_x = k_y = 150\,N\left(T/m^3\right)$$

Vertical reaction coefficients are estimated as following:

$$k_z = (1/\lambda)k_{x,y}\left(T/m^3\right)$$

Where: λ is the ratio between horizontal and vertical coefficients, which is in the range1/2÷1/5 and usually equal 3.

Comments:

According to Vietnamese standards (ex: TCVN 10304:2014 – Pile foundation – Design standard), the calculation of reaction coefficients depends on the table of constant of proportionality (5 values) and the depth of pile. The method of using the table is not completely accurate, because the reaction coefficient k depends not only on the soil type and severally it's physical mechanical properties. On the other hand, the range of values is very wide (the largest values are about 1.5÷2 times larger than the smallest). This standard does not apply to the pile foundation of coastal and offshore structures.

According to Japanese standard, it is easy to determine the reaction coefficients, reaction coefficient of transverse, vertical of pile and pile tip when no pile test result (coefficient determined based on the relationship between port structure and displacement by the equivalent method using geological survey results at the construction site). Those formulas are used to calculate reaction coefficients, they has been studied separately for the port structure and other facility structures, especially the structure of pile foundation.

In addition, the Japanese Port Standard (OCDI 2009) specifies the calculation of the reaction coefficients in the dynamic analysis that doubles the initial reaction coefficients.

In the guideline of dynamic analysis testing, the reaction coefficients of the design dynamic analysis models are triple the initial reaction coefficients.

3.3 Determine Nature Frequency by Professional Experience Method

Predict the types of natural oscillation frequencies that can appear in the measured data series (natural oscillation frequency of structure, electric, anchor line, wind …). Exclusion method is used to determine the search frequency.

At or near the natural frequencies, the phase-lag will be 90°. The search frequency of port structure is defined at 90°.

Determination of the measured natural frequency by Fast Fourier Transform of the measured acceleration range on time. This data interval is approximately zero after the moment of load impact (refer Fig. 8a).

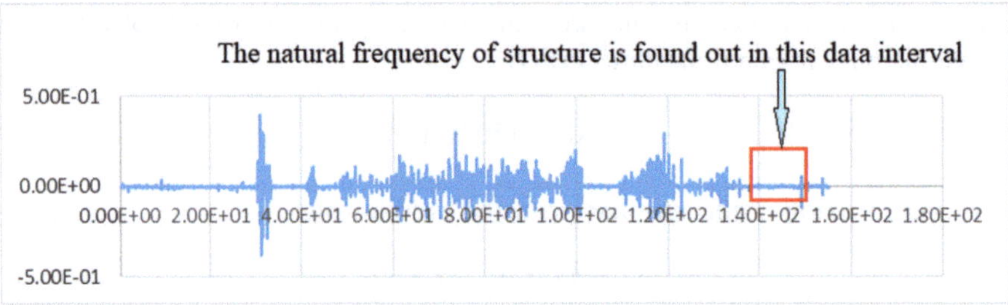

Fig. 8a. The data interval of acceleration response for the search frequency - Segment 1 of port structure

One of two software can be used to determine the natural frequency from the measured data: Fourier FFT by Mathlab software, Seismosignal software.

The determination also follows the experience of analysis experts.

3.4 Practical Application

The health assessment of port structure in Binh Thuan, Vietnam. This structure includes 3 main segments and approach trestle.

Standard Value of Natural-Vibration Frequency

Natural-vibration frequency of segments is 1.305.

Measured Natural Frequency

Segment 1:

- The measured data of acceleration response on time - Segment 1 in site (refer Fig. 8b).
- The value of measured natural frequency is 1.15967 (shown in Fig. 8c and 8d).

Segment 2:

- The measured data of acceleration response on time - Segment 2 in site (refer Fig. 9a).
- The value of measured natural frequency is 1.06812 (shown in Fig. 9b).

Segment 3:

- The measured data of acceleration response on time -Segment 3 in site (refer Fig. 10a).
- The value of measured natural frequency is 1,15967 (shown in Fig. 10b)

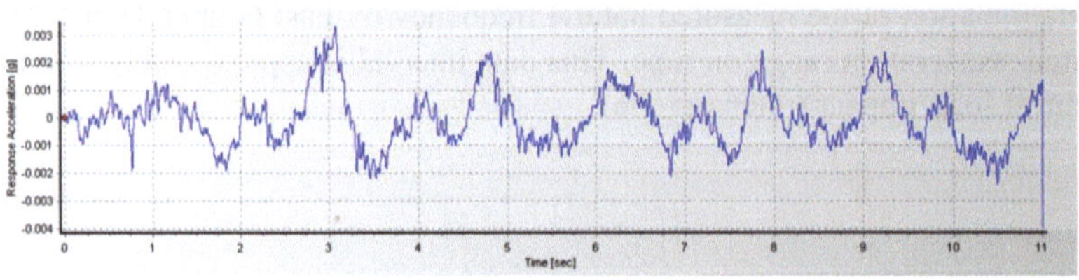

Fig. 8b. The graph of acceleration response on time - Segment 1

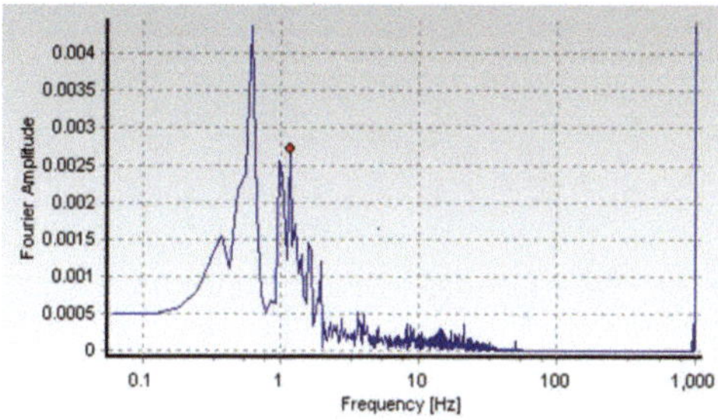

Fig. 8c. Fourier amplitude spectrum in frequency domain

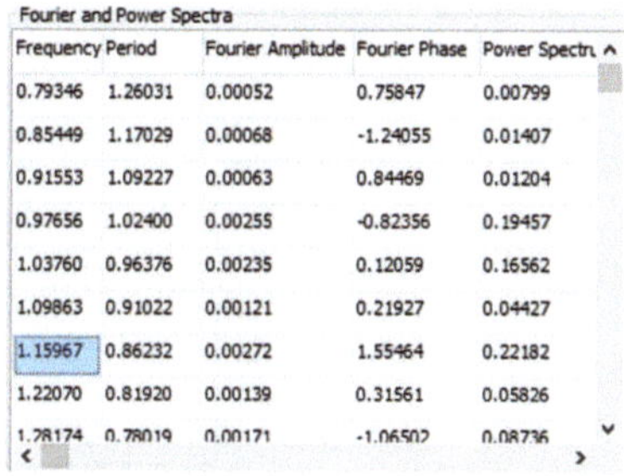
Fourier and Power Spectra

Frequency	Period	Fourier Amplitude	Fourier Phase	Power Spectr.
0.79346	1.26031	0.00052	0.75847	0.00799
0.85449	1.17029	0.00068	-1.24055	0.01407
0.91553	1.09227	0.00063	0.84469	0.01204
0.97656	1.02400	0.00255	-0.82356	0.19457
1.03760	0.96376	0.00235	0.12059	0.16562
1.09863	0.91022	0.00121	0.21927	0.04427
1.15967	0.86232	0.00272	1.55464	0.22182
1.22070	0.81920	0.00139	0.31561	0.05826
1.28174	0.78019	0.00171	-1.06502	0.08736

Fig. 8d. Determination of natural frequency based on phase angle

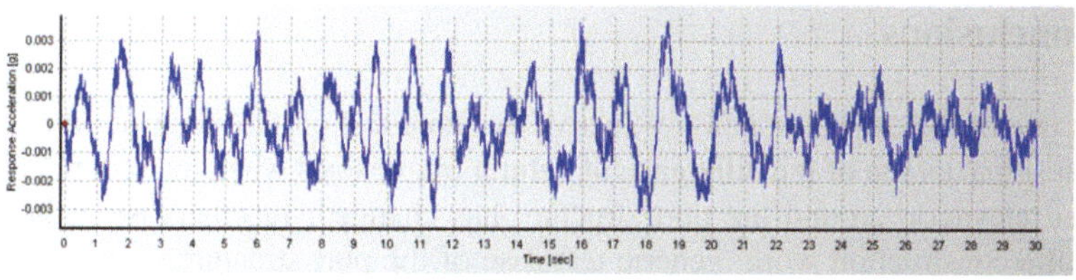

Fig. 9a. The graph of acceleration response on time - Segment 2

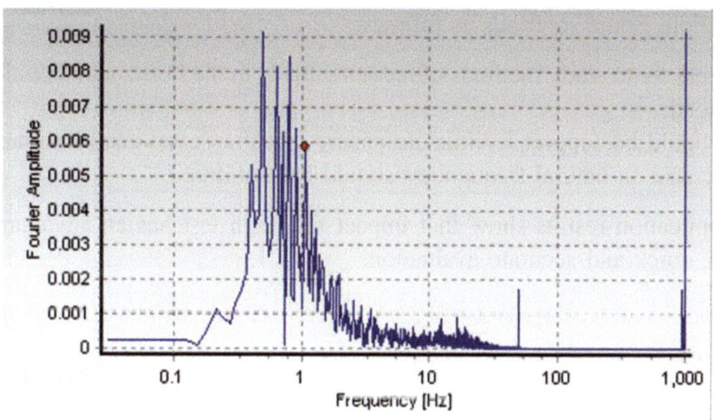

Fig. 9b. Fourier amplitude spectrum in frequency domain (1.06812)

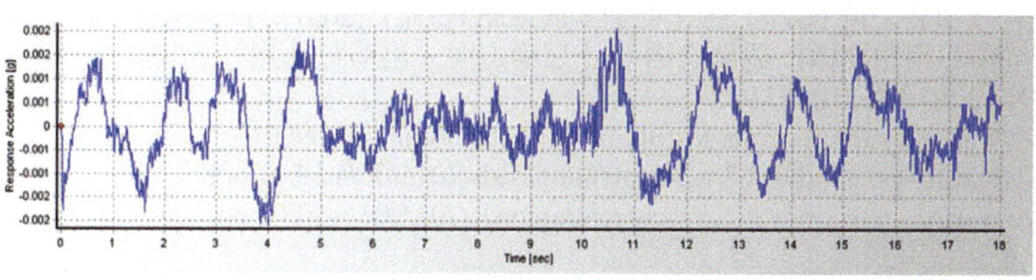

Fig. 10a. The graph of acceleration response on time - Segment 3

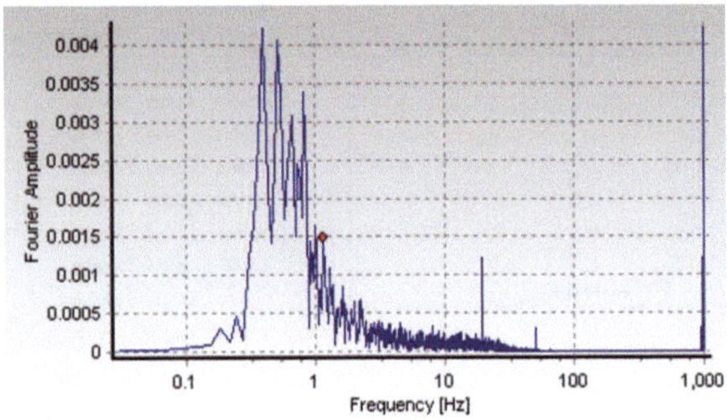

Fig. 10b. Fourier amplitude spectrum in frequency domain (1.15967)

4 Conclusions

IVT is one of the utilizable non-destructive dynamic test on the scour of port structure as the test makes use of the difference of natural frequencies when the scour occurs and riverbed elevation around the port structure has changed. In otherwise, it is used to assess the construction work, generic test case of the port structure.

Application of the test method for substructure evaluation of one new port structure in Binh Thuan, Vietnam is introduced as a case study. The health index of structure κ is 0.818. So, the health assessment of this structure is A2 level. It shows a few problems in construction. Some solutions as follows:

- It is necessary to study methods of reducing loads in the horizontal direction and the vessel load;
- There should be support facilities to ensure that the vessel can safely anchor;
- The decrease in natural frequencies should be monitored.

The application results show that impact vibration test has an advantage of non-destructive, quick and accurate evaluation.

References

Ports and Harbours Bureau, Ministry of Land, Infrastructure, Transport and Tourism (MLIT) National Institute for Land and Infrastructure Management, MLIT Port and Airport Research Institute: Technical Standards and Commentaries for port & Harbour port Facilities. The Overseas Coastal Area Development Institute of Japan, Tokyo, Japan (2009)

鉄道総合技術研究所 (編集), 国土交通省鉄道局 (監修) (2007). 鉄道構造物等維持管理標準·同解説(構造物編) 礎構造物·杭土圧構造物. 出版社 丸善, ISBN-10: 4621077724, ISBN-13: 978-4621077726

http://www.m-netis.mlit.go.jp/index.php?d=&p=detail&id=CB-090013

http://www.earth-po.co.jp/web2/wp-content/uploads/2015/05/nsirb01.pdf

http://www.earth-po.co.jp/web2/wp-content/uploads/2015/05/nsirb02.pdf

Cyclic Thermal Effects on Soil-Structure Interaction in Case of Energy Piles

Rajni Saggu(✉)

Department of Civil Engineering, Galgotias College of Engineering and Technology, Greater Noida 201306, India
rajnirupal@gmail.com

Abstract. This paper presents the effect of thermal and mechanical load on stress-strain response of energy piles in Ottawa Sand. The study is focussed on the thermally induced displacement of energy piles due to cyclic thermal loading. The pile response to combined thermal and mechanical loading is analysed through axisymmetric nonlinear finite element analysis procedure using Abaqus software for a single energy pile in Ottawa sand. The stress-strain response of pile is assumed as linear elastic and the stress-strain response of sand is simulated using constitutive model CASM. The CASM model is incorporated in finite element software Abaqus through user defined material subroutine. The pile base displacement-time history and shear stress in soil at pile-soil interface are analysed. The analyses are performed for 50 thermal cycles considering different pile base conditions (i) floating base pile and (ii) fixed base pile. A parametric study is carried out by varying (i) stiffness of surrounding soil and (ii) temperature change applied on the pile. It is concluded from the results that the pile response to thermomechanical loading can be attributed to stiffness of soil and the magnitude of thermal load applied on the pile. The mechanical loading of the piles causes downward displacement and thermal load causes upward displacement of pile head. Consequently, the shear stress response at pile-soil interface is altered. The negative shear stress generates in the fixed base pile in loose sand. The resultant pile displacement is controlled by the magnitude of applied thermal load. Elastic behaviour of piles is observed for the piles subjected to cyclic thermomechanical loading.

1 Introduction

The technology of heat exchange through the structural piles have been successfully adopted in many countries e.g. United States, Australia, Austria, Switzerland and United Kingdom (Laloui et al. 2006, Brandl 2006, De Moel et al. 2010). Even though, the studies regarding alteration in pile-soil interaction as a result of cyclic thermal load on the energy piles are rather limited in the present literature. A reinforced concrete structural pile can be used as a heat exchanger between the building and the supporting soil by installation of high density polyethylene pipes in the pile shaft which circulate a heat carrier fluid through them. In the recent literature, the experimental investigations on thermomechanical response of energy piles have been reported by Laloui et al. (2006), Bourne-Webb et al. (2009), Wang et al. (2013) and Murphy et al. (2014).

D. Choudhury et al. (Eds.): GeoMEast 2018, SUCI, pp. 183–194, 2019.
https://doi.org/10.1007/978-3-030-01920-4_16

Laloui et al. (2006) have noted heating induced expansion of a pile which results in development of compressive strain and load on the pile. Bourne-Webb et al. (2009) have reported that at the pile-soil interface, a temperature increase leads to the increase in shaft resistance mobilized by axial load. According to Wang et al. (2013), the pile shaft resistance increases as a consequence of heating of the pile installed in unsaturated dense sand. Murphy et al. (2014) concluded that the thermally induced stresses and strains in the pile increase with increase in temperature change. The centrifuge tests have been presented to investigate the pile response by Wang et al. (2014), Stewart and Mc Cartney 2014 and Goode et al. (2014). Wang et al. (2014) have reported the influence of temperature on the load carrying capacity of energy piles. Stewart and Mc Cartney (2014) concluded that the thermal strains generate consistently negative (expansive) strains on the pile head during heating of the pile under constant applied thermal load. Goode et al. (2014) have reported an initial settlement of pile head due to axial load and then upward displacement of pile head as a result of heating of the pile. Among the numerical studies, Peron et al. (2011), Di Donna and Laloui (2013) and Saggu and Chakraborty (2015a, 2015b, 2016a, 2016b and 2017) have performed numerical analyses in order to understand the thermomechanical behaviour of energy piles and reported thermally induced pile displacement and axial stresses in the piles. Peron et al. (2011) discussed the interaction between pile and soil upon thermal loading of pile and suggested that the generation of negative shear stress at the pile-soil interface should be the decisive factors in the design of energy piles. Di Donna and Laloui (2013) have reported the elastic thermal displacement of the energy pile under cyclic thermal loading. Saggu and Chakraborty (2015a) have concluded that magnitude of temperature change on the pile plays a key role in deciding the pile response. Saggu and Chakraborty (2015b) investigated the pile-soil interaction under cyclic thermomechanical loading of the piles using Mohr-Coulomb model and reported insignificant change in pile shaft capacity and significant generation of axial stresses in the energy piles. Saggu and Chakraborty (2016a) have reported the influence of shape and size of the sand particles on shear stress magnitude in soil at pile-soil interface. In another study, Saggu and Chakraborty (2017) have stated that the longer piles exhibit higher thermal expansion and the axial stresses may exceed the allowable limit under severe thermal loading conditions. In summary, the energy pile behaviour is analysed by many researchers for a single thermal cycle applied on the pile. Saggu and Chakraborty (2016b) have reported differential settlement of piles in a group consisting of mechanical and thermal piles placed altogether. They have also noted that the axial stress is higher in thermal piles than in the mechanical piles. In actual practice, the energy piles experience cycles of heating and cooling and however, rigorous finite element (FE) analysis based predictions on cyclic thermomechanical response are rare in the present literature. The objective of the present study is to determine the thermally induced pile displacement and the corresponding shear stresses in the soil at pile-soil interface for the piles subjected to multiple cycles of heating and cooling. In order to accomplish this objective, axisymmetric nonlinear finite element analyses of energy piles are performed herein for the piles subjected to 50 thermal cycles using the FE software Abaqus (Abaqus Manual Version 6.11). The investigations are performed by varying (i) the relative density (D_R) of soil and (ii) thermal load, ΔT applied on the pile. The pile is assumed to behave linear elastically and the soil behaviour is simulated with constitutive model CASM (Yu 2006) which is based on the concepts of critical state soil mechanics. The constitutive model is incorporated in Abaqus software through user defined material subroutines (i) UMAT for defining the mechanical stress-strain response of soil and (ii) UMATHT for defining the thermal parameters of soil. A uniform temperature change is applied over the entire length of the pile (Laloui et al. 2006).

2 Finite Element Simulation and Material Properties

In the present study, axisymmetric finite element analyses are performed using Abaqus software to understand the soil-structure interaction during cyclic thermal loading and to predict the pile displacement and shear stresses in soil. A concrete pile is modelled as linear elastic material having elasticity modulus $(E) = 33\,\text{GPa}$ and Poisson's ratio $(\nu) = 0.2$. The concrete thermal parameters are considered as thermal conductivity $(k_c) = 2.1\,\text{W/m}^{\circ}\text{C}$, specific heat capacity $(C_c) = 800\,\text{Joule}/^{\circ}\text{C}$ and coefficient of thermal expansion $(\alpha_c) = 10^{-5}/^{\circ}\text{C}$. The analyses are performed for pile length $(L_P) =$ 15 m and pile diameter $(B) = 0.5\,\text{m}$. For all the analyses, the soil surrounding the pile is considered as Ottawa sand. The Ottawa sand is modelled with constitutive model CASM. The vertical soil boundary is placed at a distance of $15B$ from the pile-soil interface and the bottom soil boundary is placed at a distance equal to $1L_P$ from the bottom of the pile. The vertical soil boundary is restrained in horizontal direction and the bottom soil boundary is restrained in both vertical and horizontal directions. The thermal boundary conditions include heat flow through the far-field side and the bottom boundaries of the mesh. A constant temperature of 15 °C is applied at the top soil surface. Heat flux is considered as zero along the axis of symmetry. The initial ground temperature is assumed to be 15 °C (Laloui et al. 2006). The analysis domain is meshed using four node axisymmetric thermally coupled quadrilateral, bilinear displacement and temperature elements (CAX4T) for the pile and the soil. A refined mesh with minimum element size of 0.075 mm is used for soil near the pile-soil interface. The maximum element size of 0.15 mm is used near the far-field regions of soil domain. Figure 1 shows a typical mesh for pile and soil domain. The Ottawa sand used herein is designated as ASTM C778 with $D_{50} = 0.39\,\text{mm}$. To model the interface between the pile and the soil surface, a frictional contact with coefficient of friction between concrete and soil is taken as 0.55 considering $\mu = tan(\phi_c)$ (Loukidis 2006) in the tangential direction and a hard contact with zero penetration is considered in the normal direction, where ϕ_c is critical state friction angle of soil. A gap conductance value equal to 0.274 W/m °C (thermal conductivity of Ottawa sand) is defined at the interface of pile and soil. For determination of CASM model parameters of Ottawa sand, a FORTRAN code for drained and undrained triaxial compression test simulation is written. The model parameters have been obtained for Ottawa sand with minimum and maximum void ratio of $e_{min} = 0.48$ and $e_{max} = 0.78$ (Murthy et al. 2006), respectively. The soil thermal parameters are obtained from Tarnawski et al. (2009). CASM and thermal parameter of soil are presented in Table 1. The thermal constitutive model is implemented in Abaqus through user defined material subroutine UMATHT.

The thermal response of pile-soil system is assumed to follow the energy balance equation (Green and Naghdi 1992) given by

$$\int_V \rho \dot{U} dV = \int_V r dV + \int_S q_H dS \tag{1}$$

where V is a volume of solid material, with surface area S; ρ is the density of the material; $\dot{U}$ is the material time rate of the internal energy; q_H is the heat flux per unit area of the body, and r is the heat supplied externally into the body per unit volume. The thermal constitutive relations written in terms of k and C are given by

$$\frac{dU}{dT} = C_s \tag{2}$$

and

$$\frac{dq_H}{\partial T / \partial x} = -k_s \tag{3}$$

where x is the position coordinate of any point.

The finite element analyses are carried out for thermo-mechanical loading of piles. The analyses are performed in two steps; (i) a static step to apply the gravity load to simulate the geostatic equilibrium condition of the model and (ii) a coupled temperature-displacement step to apply thermal as well as mechanical load. A thermal load (ΔT) of 21 °C with heating period of 12 days and cooling period of 16 days is applied uniformly along the pile length, similarly as applied by Laloui et al. (2006). The analyses are performed for 50 thermal cycles. The initial ground temperature is assumed as 15 °C. An allowable axial load magnitude of Q = 800kN is applied on the pile head. In present analyses, the allowable axial load has been decided by considering a factor of safety as 2.5 over the ultimate load corresponding to pile base settlement of 10% of pile diameter. The analyses are performed for soil relative densities $D_R = 50\%$ (loose sand) and $D_R = 70\%$ (dense sand). To analyse the effect of pile end conditions on thermomechanical response of the piles, floating base pile in a single layer of Ottawa sand and fixed base pile with a stiff soil layer at the pile base and Ottawa sand up to pile depth are simulated. To consider the isotropic in situ stress conditions, the coefficient of lateral earth pressure at rest (K_0) is assumed to be 1. For all the analyses performed herein, thermally induced expansive strain in the pile, tensile strain in the pile, tensile axial load and stress and upward shaft resistance are considered as positive; depth is considered positive in the downward direction from the ground surface. Displacement is considered positive in the upward direction.

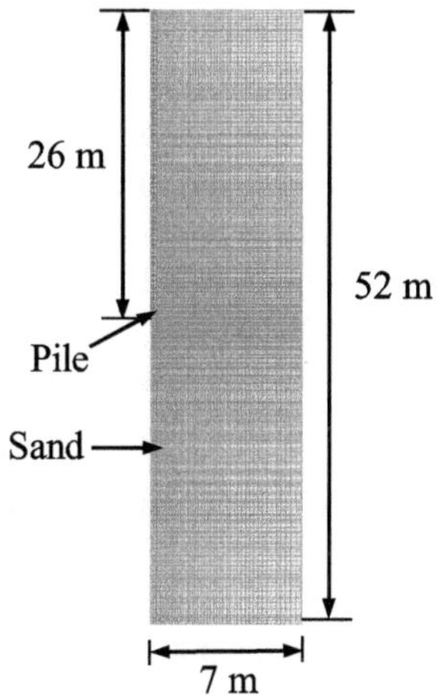

Fig. 1. Typical mesh for pile and soil domain

Table 1. Material parameters used for Ottawa sand

CASM model and thermal parameters for Ottawa sand		
Mechanical parameters	Value	Reference
Slope of critical state line in e-ln((p')) space, λ	0.02	Loukidis (2006)
Slope of unloading-reloading line in e-ln(p') space, κ	0.005	Yu (2006)
Critical state line intercept at 1 kPa pressure, Γ	0.8	Model simulation
Reference state parameter, ξ_R	0.075	Yu (2006)
Model parameter, n	4.5	Model simulation
Slope of critical state line in $q-(p')$ space, M_{cc}	1.2	Yu (2006), Loukidis (2006)
Poisson's ratio, υ_s	0.3	Yu (2006)
Shear modulus, G	32.6 MPa	Yu (2006), Murthy et al. (2006)
Bulk modulus, K_b	15 MPa	
Thermal conductivity, k_s	0.274 W/m°C	Tarnawski et al. (2009)
Specific heat, C_s	722 J/°C	
Coefficient of thermal expansion, α_s	10^{-4}/°C	

3 Validation of Finite Element Model

The validity of thermomechanical analysis of energy pile using CASM is ensured by comparing the numerical simulation results with the field pile load test data and the results of numerical simulations performed by Laloui et al. (2006). Laloui et al. (2006) performed experimental investigation on end bearing energy pile installed in layered soil consisting of alluvial soil, sandy gravelly moraine and mollase. They performed thermal pile load test for a thermal load of $\Delta T = 21\ ^{\circ}\mathrm{C}$ for a heating period of 12 days and a cooling period of 16 days and performed numerical analysis of energy pile using Drucker-Prager model for soil layers. The elastic and thermal material properties for the concrete pile and the soil layers have been taken similar to Laloui et al. (2006). The parameters for the CASM model λ, κ, Γ, ξ_R and n are assumed same as that of Ottawa sand. Figure 2 shows that the vertical stress along the pile depth obtained from present simulation is comparable with reasonable accuracy with the field pile load test data and the numerical simulation results presented by Laloui et al. (2006).

4 Cyclic Thermo-Mechanical Loading of the Pile

Energy piles undergo multiple cycles of heating and cooling during their lifetime. Therefore, it becomes necessary to study the pile-soil interaction for a number of cycles of cooling and heating operations. A set of analyses have been carried out for thermal heating-cooling cycle of 28 days frequency (12 days of heating and 16 days of cooling) and total 50 thermal cycles applied on the pile. Cyclic thermal analyses have been performed on a pile with $L_P = 15$ m and $B = 0.5$ m. The soil around the pile have been considered in isotropic in-situ stress condition with $K_0 = 1$ and the piles are considered to be installed in sand with (i) $D_R = 50\%$ and (ii) $D_R = 70\%$. Parametric sensitivity study have been carried out for cyclic thermo-mechanical loading for temperature changes (i) $\Delta T = 21^{\circ}\mathrm{C}$ and (ii) $\Delta T = 35\ ^{\circ}\mathrm{C}$ applied on the pile. A maintained axial load of 800 kN is applied at the pile head.

5 Results and Discussions

5.1 Pile Displacement

(i) *Floating base pile*

Figure 3(a) shows the effect of variation in temperature change on the base displacement of the floating base piles in sand with (i) $D_R = 50\%$ and (ii) $D_R = 70\%$, when subjected to cyclic thermo-mechanical loading. The results have been compared for temperature changes, $\Delta T = 21\ ^{\circ}\mathrm{C}$ and $\Delta T = 35\ ^{\circ}\mathrm{C}$ applied on the pile. For pile in sand with $D_R = 50\%$, high base displacement occurs as compared to the pile in sand with $D_R = 70\%$ for $\Delta T = 21^{\circ}\mathrm{C}$; which may be attributed to the higher thermal expansion of the pile at pile base in loose soil. For $\Delta T = 35\ ^{\circ}\mathrm{C}$, the base displacement for the pile in sand with $D_R = 70\%$ becomes higher which may be attributed to the restricted radial expansion offered by stiff soil near pile-soil interface. Higher temperature change causes higher base displacement of the floating pile. Thus, the increase in temperature applied on the pile results in increase in axial expansion of the pile; consequently, the pile base displacement increases.

(ii) *Fixed base pile*

Figure 3(b) shows the effect of variation in temperature change on the base displacement of the fixed base piles in sand with (i) $D_R = 50\%$ and (ii) $D_R = 70\%$, when subjected to cyclic thermo-mechanical loading. The results have been compared for temperature changes, $\Delta T = 21\ ^{\circ}C$ and $\Delta T = 35\ ^{\circ}C$ applied on the pile. For the end bearing piles, the base displacement is not affected significantly by the temperature change.

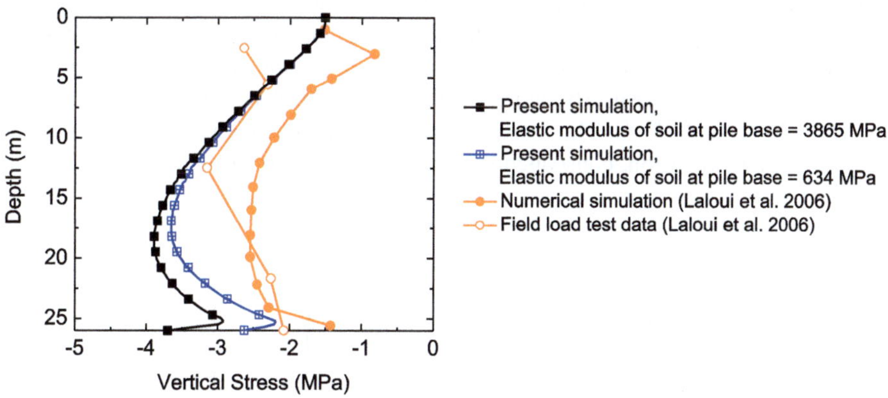

Fig. 2. Comparison of vertical stress along the pile depth for validation of thermomechanical loading with results presented by Laloui et al. (2006).

In order to understand the displacement response of the piles under cyclic thermo-mechanical loading and subjected to different magnitudes of temperature change, the displacement time history of pile nodes at the pile-soil interface have been plotted at 5 m intervals starting from the pile head as shown in Figs. 4(a) through (d). For the floating base and fixed base piles in sand with $D_R = 50\%$ and subjected to a temperature change of $\Delta T = 35\ ^{\circ}C$, the upward expansion of the pile along the upper half of pile increases towards the pile head and the downward expansion of the pile along the lower half of the pile increases towards the pile base. However, magnitude of the upward and downward movement of the pile is same under the cycles of heating. Pile displacement due to cooling of pile is also same under the cooling cycles. It can be concluded that the initial settlement of the pile occurs due to mechanical axial load at the pile head; heave at pile head and settlement at pile base happens during each cycle of heating. This implies that the pile behaviour is elastic under cyclic heating and cooling of the pile. For the fixed base piles in loose and dense sands, an increase in temperature change on the pile causes higher displacement at the pile head.

5.2 Shear Stress in Soil at Pile-Soil Interface

(i) *Floating base pile*

Figure 5(a) shows the shear stress along the pile depth at pile-soil interface for the floating base piles in sand with (i) $D_R = 50\%$ and (ii) $D_R = 70\%$. Results have been compared for two different loading cases; (i) mechanical loading and 1st heating of the pile and (ii) mechanical loading and 50th heating of the pile. It is noted that the heating of the pile causes the expansion of the pile; upward expansion near the pile head and downward expansion at the pile base. Hence the shear stress at the pile-soil interface is negative near the top 3 m length of the pile and for the remaining length of the pile, shear stress is positive. However, the shear stress magnitudes at the end of 1st heating and at the end of 50th heating are nearly same. The parametric sensitivity studies are performed for $\Delta T = 21$ °C and 35 °C. The increase in temperature change on the pile causes an increase in pile expansion; consequently increasing the shear stress magnitude.

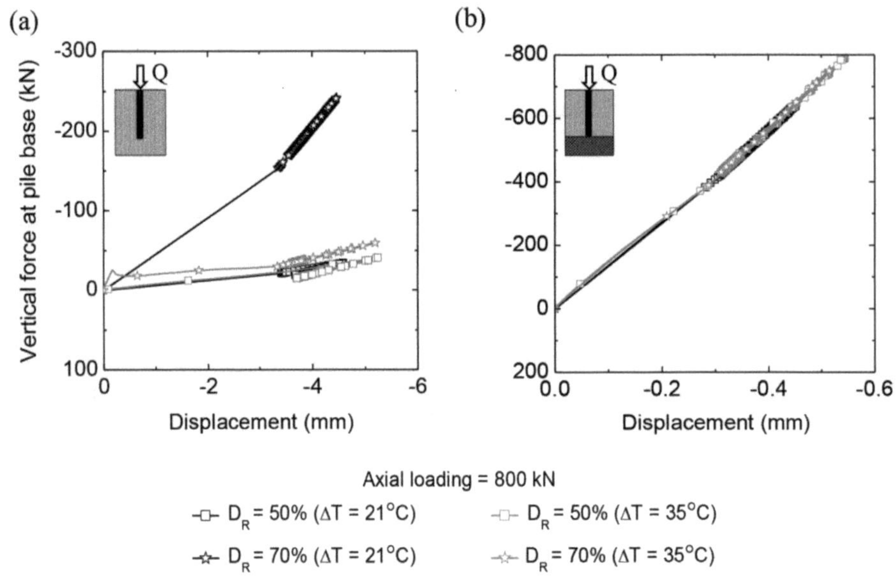

Fig. 3. Pile base load-displacement plots under combined mechanical-cyclic thermal loading under different thermal loads for (a) floating base piles and (b) fixed base piles.

(ii) *Fixed base pile*

Figure 5(b) shows the shear stress along the pile depth at pile-soil interface for the fixed base piles in sand with (i) $D_R = 50\%$ and (ii) $D_R = 70\%$. Results have been compared for two different loading cases; (i) mechanical loading and 1st heating and (ii) mechanical loading and 50th heating. Upward expansion of the pile is observed for all the cases as the downward expansion of the piles is restrained due to the presence of stiff bearing layer at the pile base. An increased temperature change from $\Delta T = 21\ ^{\circ}$C to $\Delta T = 35\ ^{\circ}$C results in significant increase in negative shear stress magnitude. However, a small magnitude of positive shear stress develops for all the piles for 5 m length from the pile base as the piles push into the soil and compress the soil at the base.

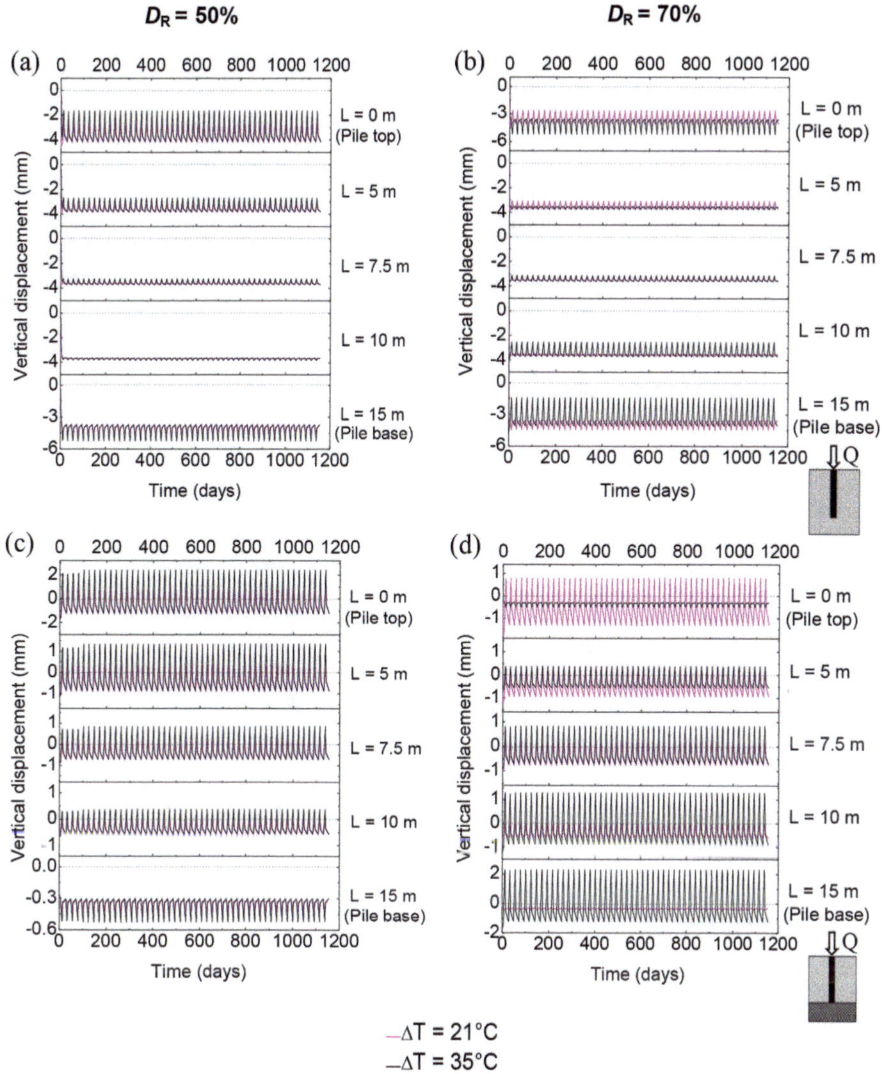

Fig. 4. Displacement-time history plots of pile nodes under combined mechanical-cyclic thermal loading and subjected to different magnitudes of temperature change for floating base and fixed base piles in sand with different relative densities.

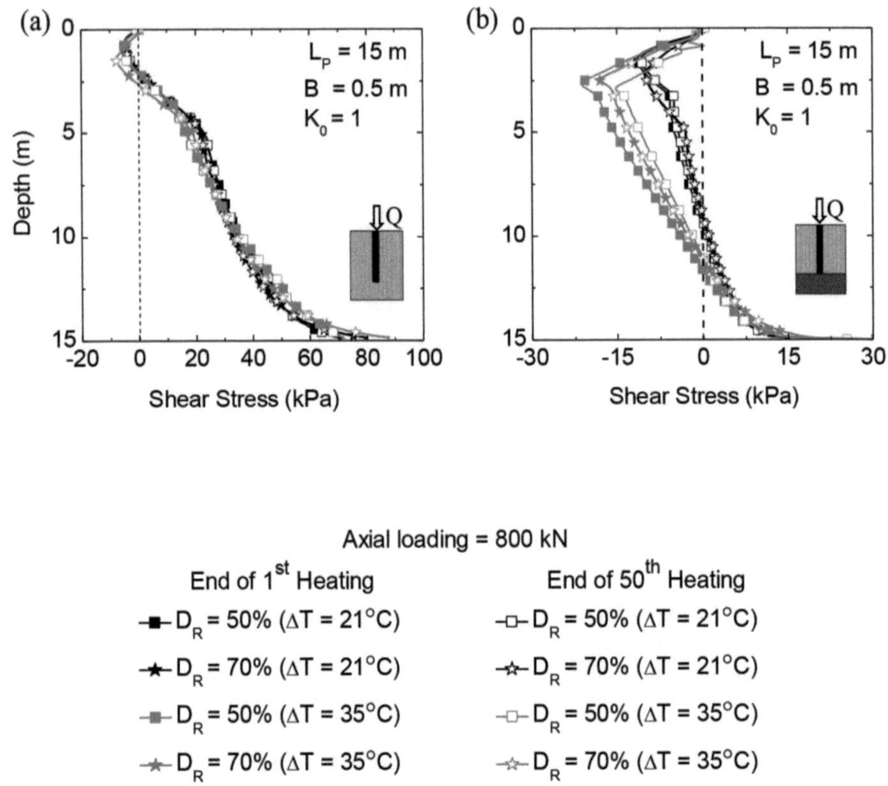

Fig. 5. Shear stress at pile-soil interface under combined mechanical-cyclic thermal loading (a) floating base piles and (b) fixed base piles in sand with different relative densities.

6 Conclusions

Thermo-mechanical response of energy piles in sand has been investigated using axisymmetric nonlinear finite element analysis procedure. Axially loaded piles under constant allowable axial load and 50 cycles of thermal loading-unloading (heating-cooling) condition have been analysed herein using the finite element code Abaqus. Critical state model, CASM is used for simulation of soil response through user defined material subroutines, UMAT and UMATHT. The response of energy piles under combined thermo-mechanical loading, e.g. (i) pile displacement and (ii) shear stress at pile-soil interface have been studied. The following conclusions are drawn from the studies conducted herein:

1. The piles show downward displacement under mechanical load and upward displacement under thermal load. Depending upon the magnitude of the thermal load, the resultant displacement of the pile head and base is controlled.
2. The end bearing piles in loose sand exhibit higher axial expansion close to pile bottom due to poor load transfer from pile to soil.

3. The relative displacement between pile and soil affects the shear stress response at the pile-soil interface. Significant amount of negative shear stress generates at the interface under thermal loading for end bearing piles in loose sand.
4. The piles subjected to cyclic thermo-mechanical loading exhibit elastic behaviour during each cycle of heating and cooling.
5. In the case of floating base piles, the shear stress does not show any significant change in magnitude at the end of 1st cycle and at the end of 50th cycle for the cases considered herein. However, in the case of end bearing piles, the shear stress magnitude decrease after 50 cycles of thermal loading.

References

Abaqus/Standard User's Manual, Version 6.11: Dassault Systèmes Simulia Corporation, Providence, Rhode Island, USA (2011)

Bourne-Webb, P.J., Amatya, B., Soga, K., Amis, T., Davidson, C., Payne, P.: Energy pile test at Lambeth College, London: geotechnical and thermodynamic aspects of pile response to heat cycles. Géotechnique **59**(3), 237–248 (2009). https://doi.org/10.1680/geot.2009.59.3.237

Brandl, H.: Energy foundations and other thermo-active ground structures. Géotechnique **56**(2), 81–122 (2006)

Di Donna, A., Laloui, L.: Advancements in geotechnical design of energy piles. In: International Workshop on Geomechanics and Energy - The Ground as Energy Source and Storage, Lausanne, Switzerland, 26–28 November 2013 (2013). https://doi.org/10.3997/2214-4609.20131946

Green, A.E., Naghdi, P.M.: On undamped heat waves in an elastic solid. J. Therm. Stresses **15**, 253–264 (1992)

Goode, J.C. III, Zhang, M., McCartney, J.S.: Centrifuge modeling of energy foundations in sand. In: Gaudin, C., White, D. (eds.) Physical Modeling in Geotechnics: Proceedings of the 8th International Conference on Physical Modelling in Geotechnics, Perth, Australia, 14–17 January 2014, pp. 729–736. Taylor and Francis, London (2014)

Laloui, L., Nuth, M., Vulliet, L.: Experimental and numerical investigations of the behavior of a heat exchanger pile. Int. J. Numer. Anal. Meth. Geomech. **30**, 763–781 (2006)

Loukidis, D.: Advanced constitutive modeling of sands and applications to foundation engineering. Ph.D. thesis, Purdue University (2006)

Murphy, K.D., McCartney, J.S., Henry, K.S.: Thermo-mechanical characterization of a full scale energy foundation. Soil Behav. Fundam. Innov. Geotech. Eng. 617–628 (2014). https://doi.org/10.1061/9780784413265.050

Murthy, T.G., Loukidis, D., Carraro, J.A.H., Prezzi, M., Salgado, R.: Undrained monotonic response of clean and silty sands. Géotechnique **57**(3), 273–288 (2006)

Peron, H., Knellwolf, C., Laloui, L.: A method for geotechnical design of heat exchanger piles. Geo-Frontiers, 470–479 (2011)

Saggu, R., Chakraborty, T.: Thermal analysis of energy piles in sand. Geomech. Geoeng. Int. J. **10**(1), 10–29 (2015a). https://doi.org/10.1080/17486025.2014.923586

Saggu, R., Chakraborty, T.: Cyclic thermo-mechanical analysis of energy piles in sand. Geotech. Geol. Eng. **33**(2), 321–342 (2015b). https://doi.org/10.1007/s10706-014-9798-8

Saggu, R., Chakraborty, T.: Pile-soil interaction under thermomechanical loading conditions imposed by geothermal energy piles in sand. In: Proceedings of Geochina 2016, Geotechnical Special Issue 259, 41–48 (2016a)

Saggu, R., Chakraborty, T.: Thermo-mechanical response of geothermal energy pile group in sand. Int. J. Geomech. **16**(4) (2016b). https://doi.org/10.1061/(asce)gm.1943-5622.0000567

Saggu, R., Chakraborty, T.: Thermo-mechanical response of geothermal energy piles in sand and parametric study. Int. J. Geomech. **17**(9) (2017). https://doi.org/10.1061/(asce)gm.1943-5622.0000962

Stewart, M.A., McCartney, J.S.: Centrifuge modeling of soil-structure interaction in energy foundations. J. Geotech. Geoenviron. Eng. **140**(4), 04013044 (2014). https://doi.org/10.1061/(ASCE)GT.1943-5606.0001061

Tarnawski, V.R., Momose, T., Leong, W.H., Bovesecchi, G., Coppa, P.: Thermal conductivity of standard sands. Part I. Dry state conditions. Int. J. Thermophys. **30**(3), 949–968 (2009)

Yu, H.S.: Plasticity and Geotechnics. Springer Publishers, New York (2006)

Wang, B., Bouazza, A., Singh, R.M., Barry, D.M., Haberfield, C., Chapman, G., Baycan, S.: Field investigation of geothermal energy pile: initial observations. In: Proceedings of the 18th International Conference on Soil Mechanics and Geotechnical Engineering, Paris, pp. 3415–3418 (2013)

Wang, W., Regueiro, R.A., McCartney, J.S.: Coupled axisymmetric thermo - poro-mechanical finite element analysis of an energy foundation centrifuge experiment in partially saturated silt. In: Abu-Farsakh, M., Hoyos, L. (eds.) Proceedings of GeoCongress 2014 (GSP 234), ASCE, pp. 2675–2684 (2014)

Probabilistic Analysis of Buried Pipeline Response Subjected to Fault Crossing

Suresh R. Dash(✉), Gautam S. Nair, Goutam Mondal, Sparsh Sehgal, and Rajesh Kumar

School of Infrastructure, Indian Institute of Technology Bhubaneswar, Bhubaneswar 752050, Odisha, India
srdash@iitbbs.ac.in

Abstract. Deterministic analysis aims to demonstrate that a facility is tolerant to identified faults/hazards that are within the 'design basis', thereby defining the limits of safe operation. However, it cannot address the risk and uncertainty associated with it adequately, especially when the degree of uncertainty is high, like in seismic events. Deterministic analysis and design can provide an appropriate degree of safety to a set of loading parameters, whereas, a probabilistic analysis and design can provide a rational design based on acceptable risk and reliability of the system performance. In this paper, the major variables that greatly influence the performance of buried pipelines subjected to fault crossing has been identified and their coefficient of variance were selected. Monte-Carlo simulation has been implemented to find out the pipeline response in terms of maximum strain in the pipeline. Different cases were taken by keeping coefficient of variance (COV) constant while changing the subsequent parameters individually as well as combined. For various state of risk, the probabilistic distribution of pipeline strain has been obtained and finally a factor of safety (FOS) is calculated as the ratio of characteristic mean response and deterministic mean (response), for various probability of exceedance (acceptable risk).

1 Introduction

Pipelines are considered as one of the most critical lifeline systems due to its significance in distributing life dependent supplies such as water, oil and gas etc. Any significant disruption to the performance of these systems often translates into undesirable impacts on economics and the living conditions of citizens. Therefore, maintaining the integrity of those lifelines during severe natural hazards such as earthquakes is of utmost importance. Pipelines traverse over a large geographical area and their exposure to various seismic hazards needs to be taken into account at the design stage. Though it is desirable to avoid areas subjected to ground movement, it is sometimes inevitable that the pipelines need to be routed through such vulnerable areas. Historically, the most catastrophic damages to pipelines are the ones resulting from faulting, seismic shaking and liquefaction. Ruptures or severe distortions of the pipeline are often associated with landslides, liquefaction, loss of support, fault surface rupture, or differential motion at abrupt interfaces between rock and soil. Among all these hazards,

D. Choudhury et al. (Eds.): GeoMEast 2018, SUCI, pp. 195–206, 2019.
https://doi.org/10.1007/978-3-030-01920-4_17

faulting is the most hazardous as it induces large strains in the pipeline, even though, it is confined to a small area close to fault zone. Given the critical importance of underground lifelines, evaluation of the response of buried pipelines at active fault crossings is amongst the top seismic design priorities. Figure 1 shows a typical example of failure of steel pipeline due to faulting during the 1999 Kocaeli earthquake. This paper looks at the design of pipeline due to faulting in a probabilistic framework as compared to deterministic design framework, which is presently being used.

Fig. 1. Large deformation and failure of steel pipeline due to active faulting during 1999 Kocaeli earthquake (adapted from Vazinram and Rasti 2006)

1.1 Current Design Practice for Pipelines Subjected to Faulting

The pipeline industry grossly follows strain based design approach. In this approach, the design check is carried out for a particular performance level by estimating the pipeline strain due to ultimate loading condition (Service Load x Factor of Safety) and limiting it to overall strain capacity (PRCI 2004). The pipelines are normally designed for two scenarios, (a) safety and (b) serviceability. The safety requirements of the pipelines subjected to seismic loading is considered by establishing the level of acceptable risk mainly based on (i) location of pipeline, (ii) size of population that is exposed to the impact of pipeline rupture, and (iii) environmental damage due to pipeline rupture. Whereas, the serviceability condition is that the pipeline should be able to maintain the supplying capability as much as possible, even under considerable local damage due to high intensity earthquake. The design level of seismic loading for safety and serviceability requirement is normally employed by application of a factor of safety, which is being considered from the probabilistic consideration of earthquake return period. Depending on the importance of the pipelines, they are normally divided into four classes and suitable design seismic action has been assigned to different classes as given in Table 1.

Table 1. Pipeline classification based on its importance and associated design seismic action (adapted from IITK-GSDMA 2007)

Class of pipeline	Description	Probability of exceedance in 50 years	Importance Factor (I)
Class I	1. Very essential water pipelines required to serve for post-earthquake response and intended to remain functional and operational during and following a design earthquake 2. High pressure oil and gas pipelines which are required to remain functional during and following the design earthquake 3. Pipelines which would cause extensive loss of life or a major impact on environment in case of failure or damage	2% (Return Period = 2475 years)	2.3
Class II	1. Critical water pipelines serving a large community and having significant economic impact to the community or a substantial hazard to human life and property in the event of failure 2. Medium pressure oil and gas pipelines which are vital energy serving facilities, but their service can be interrupted for a short period until minor repairs are made	5% (Return Period = 975 years)	1.5
Class III	1. Most of the water supply pipelines for ordinary use 2. Low pressure oil and gas pipelines	10% (Return Period = 475 years)	1

The design basis seismic loading (fault deformation) corresponding to class III pipeline is estimated based on site specific hazard analysis for an earthquake of 10% probability of exceedance in 50 years (return period of 475 years). Then a factor to account for other probability of exceedance values for classes I and II pipelines are estimated by multiplying an importance factor as given in Table 1. These importance factors apparently addresses the uncertainty that lies in the estimation of seismic hazard value, mainly fault displacement. However, this is a gross simplification of the uncertainty, which can be associated with multiple variables, such as, fault displacement, fault orientation, pipe to fault crossing angle, etc.

For a given site, considering a 475 year return period event, design fault displacement is evaluated. It is then multiplied by the importance factor as defined in Table 1 to convert the design fault displacement for other classes of pipeline. The design fault displacement (AD) can be estimated by available regression models for the region. Widely used Wells and Coppersmith (1994) model (Eq. 1) can also be used in case of non-availability of the region specific model.

$$log\,(AD) = -6.32 + 0.90M_w \quad (1)$$

where M_w is the moment magnitude. There have been a lot of studies on the probabilistic analysis of pipelines subjected to various seismic hazards over the years (McGuire et al. 1988; Elhmadi and O'Rourke 1989; Mashaly and Datta 1989; Datta 2000; Youngs et al. 2000; Stepp et al. 2001; Omidvar et al. 2013; Esposito et al. 2014; Melissianos et al. 2015; Cheng and Akkar 2016; Eskandari et al. 2017; Goda 2017) which provides the basis and procedure of probabilistic framework for seismic design of pipelines. In the deterministic approach, which is being followed in current design, the probabilistic input is included in form of the importance factors. These factors are assigned considering the difference in hazard scenario (*i.e.,* probability of exceedance in 50 years) for various class of pipelines; however, this only considers the uncertainty in fault displacement. However, the response parameter, *i.e.,* strain in the pipeline, is dependent on many parameters, whose values have high degree of uncertainty involved. Some of these parameters include fault displacement, fault crossing angle, soil parameters, and burial depth.

The present study, hence, focuses on finding out pipeline strain due to strike slip faulting for individual and combined effect of various input parameters with defined uncertainty level. The uncertainty parameters that is highly influential to the performance of buried pipelines have been identified and appropriate and respective coefficient of variance values were selected.

2 Probabilistic Analysis

To determine the pipeline strain while crossing a strike slip fault, numerical study has been carried out. The parameters that mainly contributes towards the strain in pipeline includes fault displacement and its orientation, soil properties and pipeline properties, which can be referred to in the next sub-heading. Among these, three parameters having high degree of uncertainty in evaluation of its mean design value, such as: (a) fault displacement, (b) cohesion of soil, and (c) burial depth have been considered for probabilistic analysis because of their high potential for development of strain in pipeline. Effect of other parameters and the uncertainty quotient of them are not studied in this paper, which may be looked at in future. Many sets of random variables were generated to obtain a number having a constant probabilistic mean for a given coefficient of variance (COV). A sensitivity study has been performed which signified that the variance in the pipeline strain distribution saturates at about 100000 number of analyses (Fig. 2). Hence, the number of steps in Monte-Carlo simulation is fixed as 100000 for the present study. For each parameter, 100000 values were generated for the analysis depending on the mean and standard deviation (obtained from coefficient of variance) assuming a normal distribution pattern. To economize the computational effort, a Monte-Carlo simulation has been done considering random values from the three distributions (Fig. 3).

Many researchers have studied the distribution of actual values from calculated mean values and proposed COV that have been used in the present study.

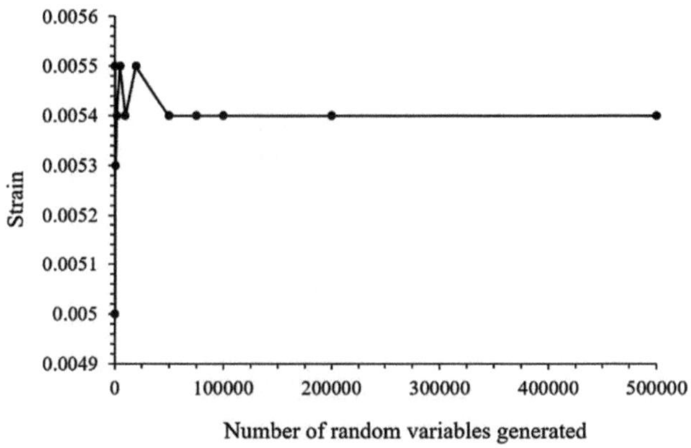

Fig. 2. Variation of strain with respect to number of random variables generated

- *Fault displacement:* The design value of fault displacement is estimated for a representative earthquake magnitude (M_w) of 6.6 for the present study, corresponding to 10% probability of exceedance in 50 years (Pipeline class III). The design fault displacement (for strike slip fault) is given by Eq. (1) (Wells and Coppersmith 1994). Associated coefficient of variance (COV) is taken as 40% ± 20% (Nicol et al. 2012).
- *Soil parameter (Cohesion, c):* For coefficient of cohesion, the coefficient of variance chosen is 40% and mean value is taken as 30000 Pa (Lizarraga and Lai 2014).
- *Burial depth (H):* The mean value of 1.2 m burial depth is used with a coefficient of variance as 5%.

For other earthquake scenarios for pipe classes II and I, the magnitude is chosen with return period of 975 years and 2475 years, respectively. These values are very much location specific. However, for the present study, which is not region specific, a representative value has been taken as M_w = 7.0 for class II and M_w = 7.5 for class I pipeline. Using these mean and standard deviation (= mean × COV) values, random variables using normal distribution were generated using the following equation.

$$f(x|\mu,\sigma) = \frac{1}{\sigma\sqrt{2\pi}} e^{-\frac{(x-\mu)^2}{2\sigma^2}} \tag{2}$$

2.1 Procedure

For any considered value from the above three variables, the strain value is calculated using the step-wise procedure as mentioned below (as per IITK-GSDMA 2007).

1. *Basic data*:
   a. Pipe diameter, D = 0.6 m
   b. Pipe thickness, t = 6.4 mm
   c. Modulus of pipe material, $E = 2 \times 10^{11}$ Pa

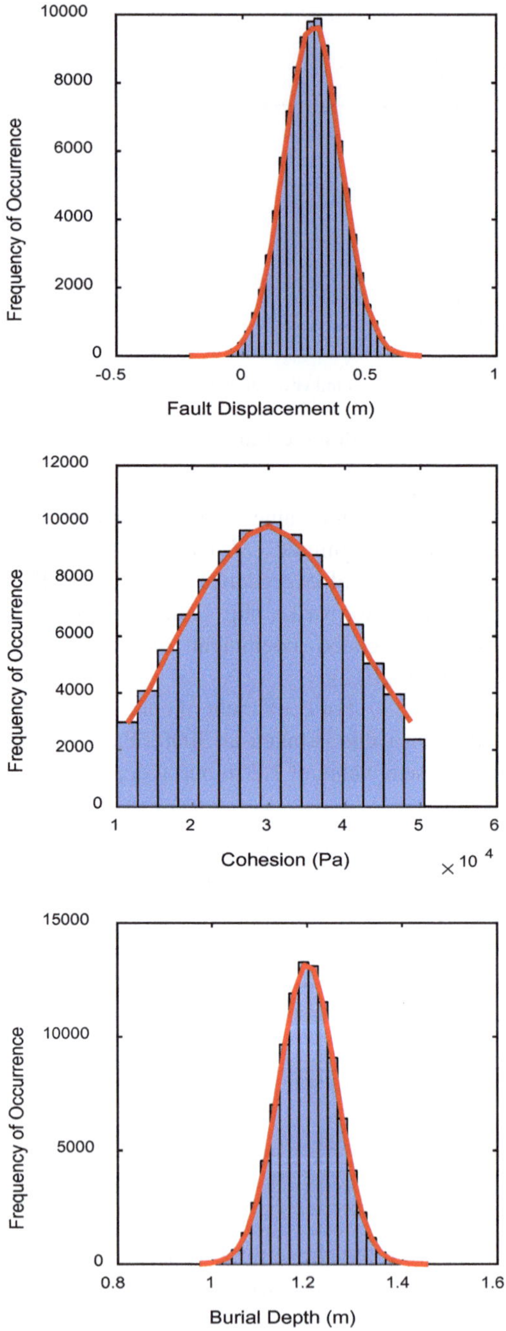

Fig. 3. Normal distribution curve for three variables considered for analysis

d. Yield strain of material, $\varepsilon_y = 0.002$
e. Effective unit weight of soil, γ = 18000 N/m^3 (for dry condition)
f. Angle of internal friction of soil, $\phi = 30°$

2. *Axial component of fault displacement*: $\delta_{fax} = \delta_{mean} \times \cos\beta$, where δ_{mean} is the mean value of fault displacement which is obtained from Table 2 and β is the pipeline – fault crossing angle = 40°

Table 2. Factor corresponding to different classes of pipeline and COV

Ratio	Coefficient of variance		
	20%	40%	60%
AD/δ_{mean} (Class III)	1.26	1.51	1.77
AD/δ_{mean} (Class II)	1.33	1.66	1.99
AD/δ_{mean} (Class I)	1.41	1.82	2.23

3. *Transverse component of fault displacement*: $\delta_{ftr} = \delta_{mean} \times \sin\beta$
4. *Adhesion factor*: $\alpha = 0.608 - 0.123c - \left(\frac{0.274}{c^2+1}\right) + \left(\frac{0.695}{c^3+1}\right)$, where c is in 100 kPa
5. *Interface angle of friction between soil and pipe*: $\delta' = f \times \phi$, where f is the friction factor = 0.7 (for smooth steel pipe)
6. *Coefficient of soil pressure at rest*: $K_o = 1 - \sin\phi$
7. *Friction force in axial direction*: $t_u = \pi D\alpha c + \pi DH\gamma\frac{1+K_o}{2}\tan\delta'$
8. *Effective unanchored length*: $L_a = \frac{E\times\varepsilon_y\times D\times t\times\pi}{t_u}$
9. *Strain due to faulting*: $\varepsilon = \left[\frac{\delta_{fax}}{2L_u} + \left(\frac{\delta_{ftr}}{2L_a}\right)^2\right]$

For all these strain calculations using different parameters having a set of random values, characteristic strain value (ε_{ch}) is obtained. The characteristic strain value is defined as the value corresponding to 95% probability of non-exceedance in 50 years (Fig. 4). This is the design strain for which the safety check is done to make it fall below permissible limit. The results obtained for various cases have been discussed in the following section.

3 Results and Discussion

The variation of characteristic strain values have been studied with respect to classes of pipelines by considering only the effect of fault displacement as well as by considering the effect of all the three parameters. The influence of COV associated with the fault displacement on the characteristic strain have also been studied. Table 3 shows the characteristic strain values obtained from the deterministic and probabilistic analyses. It can be observed that the characteristic strain value largely depends upon the classes of pipeline, the coefficient of variance associated with fault displacement, and the input parameters considered. For better understanding, the same has been plotted graphically

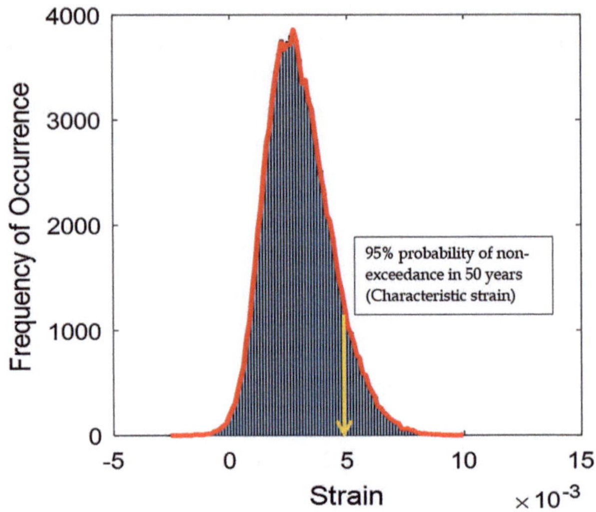

Fig. 4. Calculation of design/characteristic strain

Table 3. Characteristic strain values obtained from deterministic and probabilistic analyses

Method	Class I	Class II	Class III	Remarks	
Deterministic	0.0292	0.0103	0.0045	–	
Probabilistic	0.0275	0.0103	0.0047	Considering fault displacement only	COV (faulting) = 20%
	0.0267	0.0103	0.0049		COV (faulting) = 40%
	0.026	0.0103	0.005		COV (faulting) = 60%
	0.0317	0.0119	0.0055	Considering three parameters (*i.e.*, fault displacement, soil cohesion, and burial depth of pipe)	COV (faulting) = 20%
	0.0292	0.0114	0.0054		COV (faulting) = 40%
	0.0282	0.0111	0.0054		COV (faulting) = 60%

in Figs. 5 and 6 considering deterministic value and probabilistic value of characteristic strain corresponding to COV of 60% considering the effect of all three parameters. The difference between evaluated strain for deterministic and probabilistic analyses is plotted in the form of a ratio for the highest values obtained from the probabilistic study. It can be inferred from Fig. 6 that about 3.5% decrease and 7% increase in characteristic strain values have been observed for class I and class II pipelines, respectively, using probabilistic analysis when compared with the deterministic analysis. However, a significant increase of 20% in characteristic strain value has been observed for class III pipelines using probabilistic analysis as compared to that

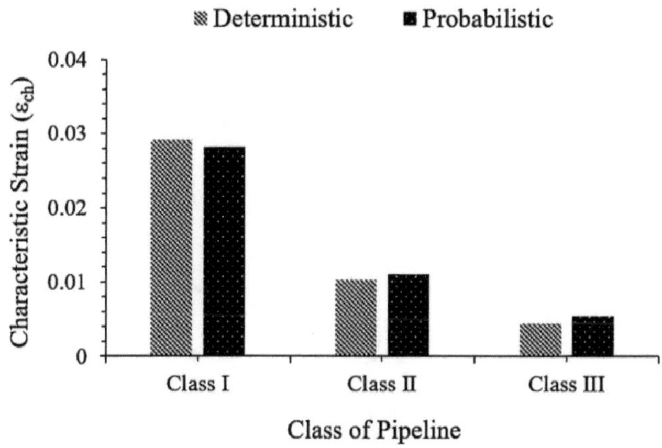

Fig. 5. Comparison of characteristic strain values with respect to different classes of pipeline using deterministic and probabilistic analyses

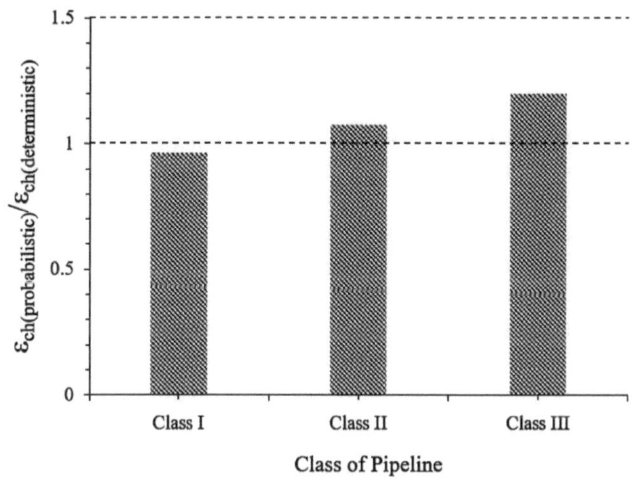

Fig. 6. Difference in characteristic strain value from deterministic and probabilistic analysis for various classes of pipelines

obtained using deterministic analysis. Figure 7 illustrates the influence of coefficient of variance on the characteristic strain value obtained using probabilistic analysis considering the effect of fault displacement only. It is observed in the analysis that as the design fault displacement (AD) remained constant for three different cases of COV for fault displacement, a decreasing trend is observed in strain values with increase in COV. Hence, critical strain values occur when COV is 20%, when other parameters are not varied.

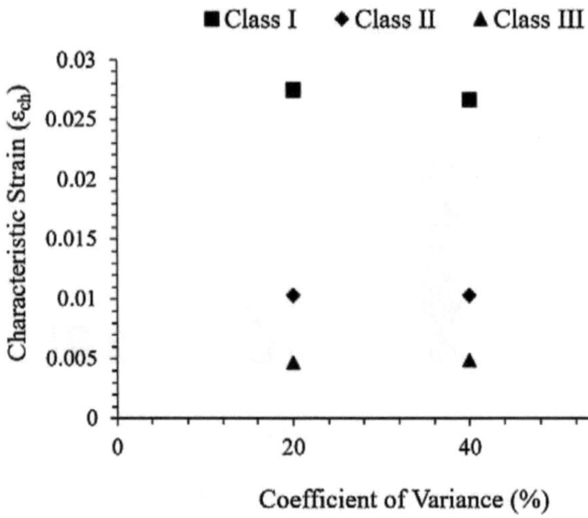

Fig. 7. Influence of COV on characteristic strain values corresponding to different classes of pipeline

4 Conclusions

The influence of major parameters such as fault displacement, soil cohesion and burial depth of pipeline on the development of strain in pipeline when subjected to strike-slip faulting have been investigated in the present study using Monte-Carlo simulation. Various cases were considered to study the individual effect of fault displacement and the combined effect of all three parameters with a defined uncertainty level. It was observed that the characteristic strain is highly dependent on the class of pipeline, coefficient of variance associated with fault displacement and the input parameters considered. The difference between evaluated strain for deterministic and probabilistic analyses was plotted for the highest values obtained from the probabilistic study. A significant increase of 20% in characteristic strain value was observed for class III pipelines, followed by a marginal increase of 7% and a decrease of 3.5% for class II and class I pipelines respectively, using probabilistic analysis as compared to that obtained using deterministic analysis. It was also observed that as the design fault displacement remained constant for three different cases of COV for fault displacement, a decreasing trend was observed in strain values with increase in COV. Hence, critical strain values occur when COV is 20%, when other parameters are not varied. Similar study can be done for other types of fault crossing to find its influence on the development of strain in pipeline employed in the design.

References

Vazinram, F., Rasti, R.: Seismic Hazards for lifelines. In: Proceedings of Geohazards, Engineering Conferences International, Norway (2006)

PRCI: Guidelines for the Seismic Design and Assessment of Natural Gas and Liquid Hydrocarbon Pipelines. Pipeline Research Council International Inc. (PRCI), R & D Forum, Canada (2004)

IITK-GSDMA: IITK-GSDMA Guidelines for Seismic Design of Buried Pipelines. Prepared by Dash, S.R., Jain, S.K. (eds.) Indian Institute of Technology Kanpur (IITK) and Gujarat State Disaster Management Authority (GSDMA), National Information Center of Earthquake Engineering (NICEE) Publication, India (2007)

Wells, D.L., Coppersmith, K.J.: New empirical relationships among magnitude, rupture length, rupture width, rupture area, and surface displacement. Bull. Seismol. Soc. Am. **84**(4), 974–1002 (1994)

McGuire, R.K.: Seismic risk to lifeline systems: critical variables and sensitivities. In: Proceedings of the 9th World Conference on Earthquake Engineering, Tokyo-Kyoto, Japan, vol. VII, pp. 129–134 (1988)

Elhmadi, K., O'Rourke, M.J.: Seismic Wave Propagation Effects on Straight Jointed Buried Pipelines. Technical Report NCEER-89-0022, National Center for Earthquake Engineering Research (NCEER), State University of New York at Buffalo, New York, USA, p. 224 (1989)

Mashaly, E.A., Datta, T.K.: Seismic risk analysis of buried pipelines. J. Transp. Eng. (1989). https://doi.org/10.1061/(asce)0733-947x(1989)115:3(232)

Datta, T.K.: Seismic response of buried pipelines: a state-of-the-art review. J. Nuclear Eng. Des. (2000). https://doi.org/10.1016/S0029-5493(99)00113-2

Youngs, R.R., Arabasz, W.J., Anderson, R.E., Ramelli, A.R., Ake, J.P., Slemmons, D.B., McCalpin, J.P., Doser, D.I., Fridrich, C.J., Swan III, F.H., Rogers, A.M., Yount, J.C., Anderson, L.W., Smith, K.D., Bruhn, R.L., Knuepfer, P.L.K., Smith, R.B., dePolo, C.M., O'Leary, D.W., Coppersmith, K.J., Pezzopane, S.K., Schwartz, D.P., Whitney, J.W., Olig, S. S., Toro, G.R.: A methodology for Probabilistic Fault Displacement Hazard Analysis (PFDHA). Earthquake Spec. (2000). https://doi.org/10.1193/1.1542891

Stepp, J.C., Wong, I., Whitney, J., Quittmeyer, R., Abrahamson, N., Toro, G., Youngs, R., Coppersmith, K., Savy, J., Sullivan, T., Yucca Mountain PSHA Project Members: Probabilistic seismic Hazard analyses for ground motions and fault displacement at Yucca Mountain. Earthquake Spec. (2001). https://doi.org/10.1193/1.1586169

Omidvar, B., Eskandari, M., Peyghaleh, E.: Seismic damage to urban areas due to failed buried fuel pipelines case study: fire following earthquake in the city of Kermanshah, Iran. Nat. Hazards (2013). https://doi.org/10.1007/s11069-013-0554-9

Esposito, S., Lervolino, L., d'Onofrio, A., Santo, A., Cavalieri, F., Franchin, P.: Simulation-based seismic risk assessment of gas distribution networks. Comput.-Aided Civ. Infrastruct. Eng. (2014). https://doi.org/10.1111/mice.12105

Melissianos, V.E., Vamvatsikos, D., Gantes, C.J.: Probabilistic assessment of innovative mitigating measures for buried steel pipeline – fault crossing. In: Proceedings of the ASME 2015 Pressure Vessels and Piping Conference, Boston, Massachusetts, USA (2015). https://doi.org/10.1115/pvp2015-45345

Cheng, Y., Akkar, S.: Probabilistic permanent fault displacement hazard via monte carlo simulation and its consideration for the probabilistic risk assessment of buried continuous steel pipelines. Earthquake Eng. Struct. Dyn. (2016). https://doi.org/10.1002/eqe.2805

Eskandari, M., Omidvar, B., Modiri, M., Nekooie, M.A., Alesheikh, A.A.: Geospatial analysis of earthquake damage probability of water pipelines due to multi-hazard failure. ISPRS Int. J. Geo-Inf. (2017). https://doi.org/10.3390/ijgi6060169

Goda, K.: Probabilistic characterization of seismic ground deformation due to tectonic fault movements. J. Soil Dyn. Earthquake Eng. (2017). https://doi.org/10.1016/j.soildyn.2017.05.039

Nicol, A., Robinson, R., Van Dissen, R.J., Harvison, A.: Variability of single event slip and recurrence intervals for large magnitude paleoearthquakes on New Zealand's active faults. GNS Science Report 2012/41, p. 66 (2012)

Lizarraga, H.S., Lai, C.G.: Effects of spatial variability of soil properties on the seismic response of an embankment dam. J. Soil Dyn. Earthquake Eng. (2014). https://doi.org/10.1016/j.soildyn.2014.03.016

A Study on the Soil Structure Interaction of a 15 Storey 2 × 3 Bay Building Subjected to Lateral Load (Earthquake Load)

S. Deepa[1,2,3,4(✉)], G. Narayan[1,2,3,4], S. V. Venkatesh[1,2,3,4], and I. R. Mithanthaya[1,2,3,4]

[1] Department of Technical Education, Government Polytechnic-Udupi, Udupi, Karnataka, India
deepasujayakumar@yahoo.co.in, narayanagbgs@gmail.com, svvenkatesh@pes.edu, Mith9999@yahoo.com
[2] Department of Civil Engineering, SJCIT, Chikballapur, Karnataka, India
[3] Department of Civil Engineering, PESIT, Bangalore, Karnataka, India
[4] Department of Civil Engineering, NMAMIT, Nitte, Karnataka, India

Abstract. The consequences of interaction between a structure and the supporting soils in a structural model including the stiffness and strength of the geotechnical (soil) components of the foundation are essential in analyzing a structure. The basic method followed by placing the structural model on a rigid base or fixed-base is not the appropriate method of analysis as it is necessary to model the structural and geotechnical (soil) components of the foundation which results in the response of the overall structural system including deformations (elastic and inelastic) in the structural and geotechnical parts of the foundation system which helps for better design. Compared with the fixed-base modeling approach, the soil structure interaction predicts period change, the distribution of forces which changes for various elements, the sequence of inelastic change and the modes of inelastic behavior. All of these result in more realistic evaluation of the structural behavior and performance. In the current study, "3D model 15 stories having [2 × 3] bay moment- resisting RC frame with Isolated and Raft foundation resting on different types of soil is analyzed. Comparison is carried out between soil flexibility (Interacting analysis) with fixed base (Non interacting analysis) for RSM method of analysis. The investigation is carried out in zone- V of seismic zones of India as per IS 1893 (part 1):2002, considering loads (dead, live and seismic loads) and their seismic load combinations". The soil is idealized by Two [2] forms of soil models i.e. Modified Winkler model and Continuum soil model. Totally 12 models are analyzed, for each approach 6 models are analyzed by RSM method for 3 types of soil conditions and 2 types of foundations. The models chosen for analysis are 1] Basic moment resisting RC frame with isolated footing and raft foundation - Modified Winkler model 2] Bare frame with isolated and raft footing - Continuum model. The results obtained analytically are precisely studied for higher values of period, base reaction, axial force, bending moment and displacement. The result suggests which soil condition requires SSI effect to be considered for analysis and also specifies which method of soil idealization gives realistic results.

D. Choudhury et al. (Eds.): GeoMEast 2018, SUCI, pp. 207–219, 2019.
https://doi.org/10.1007/978-3-030-01920-4_18

Keywords: RC frame · Earthquake load
Response spectrum method of analysis · Soil structure interaction

1 Introduction

Soils have considerable ductility unless they significantly reduce in stiffness and strength under cyclic loading. Structures are usually modeled without considering the effect of soil. The effect of soil structure interaction shall be evaluated for those buildings in which an increase in fundamental period caused by soil structure interaction results in an increase in spectral acceleration as of which for other buildings soil structure interaction effects need not be considered. SSI may modify the seismic demands on the building. It can reduce or increase spectral accelerations and seismic forces. Calculation of SSI effects using explicit modelling procedure shall be based on a mathematical model that includes flexibility and damping of individual foundation components. The foundation system shall be modelled considering the degree of fixity provided at the base of the structure. Figure 1 illustrates the assumption that the structural model is mounted on a rigid base that is excited by the free-field motion. The free-field motion is the theoretical movement of a single point on the surface of the ground, assuming that there is no structure near it. The fixed-base modeling assumption is inappropriate for many structures though structural systems that incorporate stiff vertical elements for lateral resistance can be particularly sensitive to even small base rotations and translations that are neglected with a fixed base assumption. Relatively flexible vertical elements (e.g. Moment frames) are often not significantly affected by SSI. Figure 2 illustrates the incorporation of foundation flexibility into the structural model directly. Those provisions normally use the free-field motion as the seismic demand with 5% damping as the conventional initial value. This approach is capable of modeling both the structural and geotechnical (soil) components. Compared with the fixed-base modeling approach, the predicted period of the structure lengthens the distribution of forces among various elements changes, the sequence of inelasticity and the modes of inelastic behavior can change. All of these effects result in more realistic evaluation of the probable structural behavior and performance.

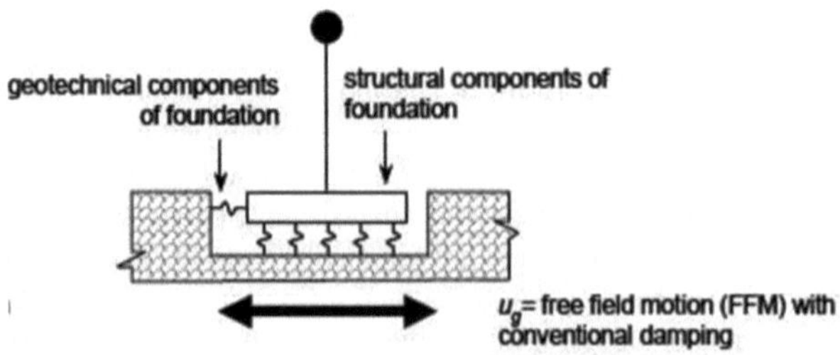

Fig. 1. Rigid base model

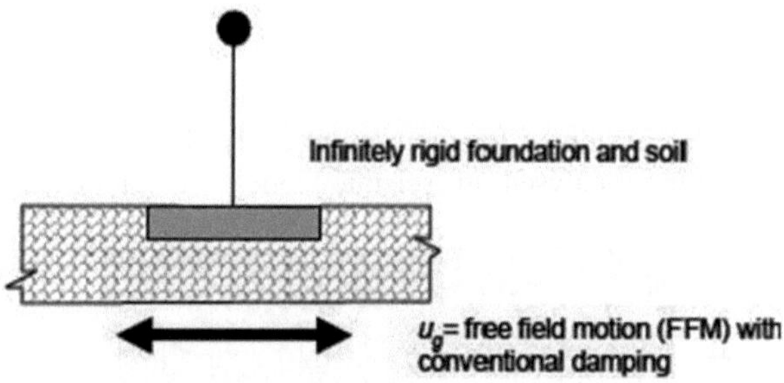

Fig. 2. Flexible base model

2 Present Investigation

3D building of 2 × 3 bays 15 storey is considered with Rigid and Flexible base. Two types of footings selected are isolated footing and raft footing and subjected to seismic analysis in zone 5 (where zone 5 is the area most prone to earthquake). Three different types of soil are considered i.e., Hard soil, Medium soil and Soft soil based on parameters Vs., N, Su which are the average values of the shear wave velocity, standard penetration test blow count, and undrained shear strength (as per standards ASCE41-13). The other parameters required are Site class, Density of soil, Poison's ratio, Specific gravity due to acceleration, Initial shear modulus, and Effective shear modulus. All these soil details along with isolated and raft footing are incorporated in 3D model where soil is idealised by two methods Modified Winkler's method and Continuum method. In this study comparison between flexible bases and rigid base for different types of soil is studied.

3 Analysis Method

The structure is modelled in SAAP 2000 V19.2; the various parameters considered for structure, footing and soil is incorporated in the model. The frame details are specified where the column and beam are taken as line element and slab and footing are taken as area element. Soil is idealised by Modified Winkler method and Continuum method the structure is considered in seismic zone 5. Response Spectrum method of analysis is performed for different types of soil conditions.

Table 1. (A): Parameters Considered (B): Stiffness solution (as per ASCE41-13)

Table 1(A): Parameters Considered

Sl.No.	Particulars	Size
1	Main beam in x-direction	0.30x 0.75m
2	Main beam in y- direction	0.30x 0.375m
3	Column	0.90x0.45m
4	Column height	3.0m
5	Slab thickness	0.125m

Table 1(B) : Stiffness solution (as per ASCE41-13)
K_{xsur}= GB/(2-γ)*(3.4*(L/B)$^\wedge$0.65+1.2)
K_{ysur}=(GB/(2-γ))*(3.4*(L/B)$^\wedge$0.65+0.4*(L/B)+0.8)
K_{zsur}= (GB/(1-γ))*(1.55*(L/B)$^{\wedge 0.75}$+0.8)
$K_{\theta xx}$=(GB^3/(1-γ))*(0.4(L/B)+0.1)
$K_{\theta yy}$=(GB^3/(1-γ))*(0.47*(L/B)$^{\wedge 2.4}$+0.034)
$K_{\theta zz}$=GB^3*(0.53*(L/B)$^{2.45}$+0.51)

3.1 Defining Problem

3D model of 2 × 3 bays, 7.5 m width in X-direction and 3 m length in Y-direction of overall height 45 m is considered for the present study. These are various parameters considered for structure analysed in seismic zone 5 is tabulated in Table 1(A).

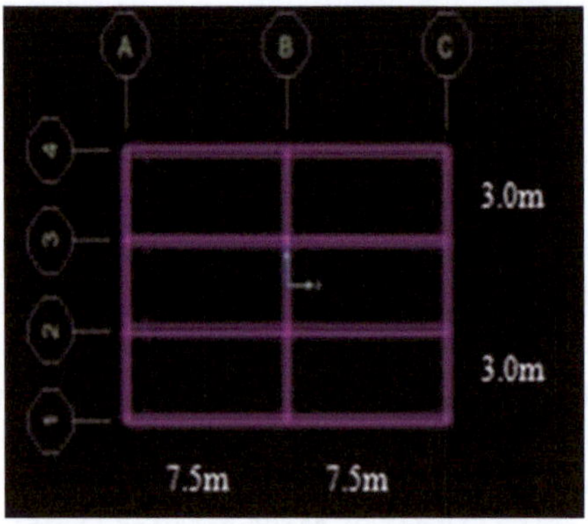

Fig. 3. Plan of bare frame (BF) (modified winkler method)

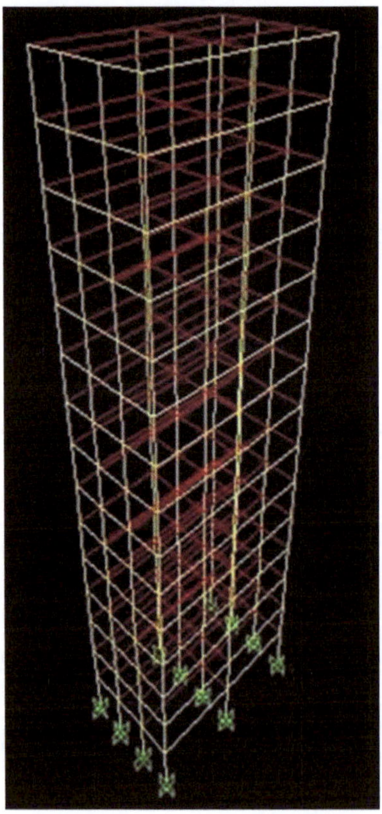

Fig. 4. Elevation of BF with fixed base (modified winkler method)

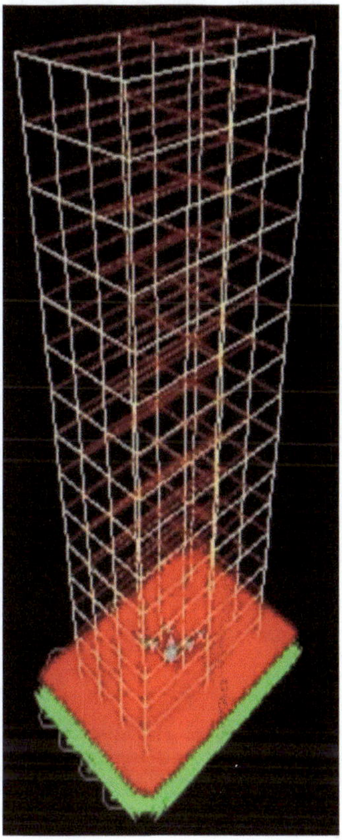

Fig. 5. Elevation of BF with raft footing (modified winkler method)

The stiffness of soil and foundation is calculated based on Stiffness solution given in ASCE 41-13 as shown in the Table 1(B) below and FEMA 440. The spring stiffness with 6 degree of freedom Kxx, Kyy, Kzz, $K_{\Theta xx}$, $K_{\Theta yy}$, and $K_{\Theta zz}$ is placed at every joint for raft footing and isolated footing. In Continuum method the soil is considered as isotropic, homogenous elastic half space (3D) for which dynamic shear modulus and Poisson's ratio taken as the input. Soil is considered around and beneath the footing, up to a predetermined zone. Soil is idealized as finite element with eight nodded elements (Solid) with three degrees of freedom of translation in the respective co-ordinate directions at each node. In addition, the structural systems are subjected to 13 different load combinations as per provision of IS 1893:2002. The Plan of Bare frame shown in

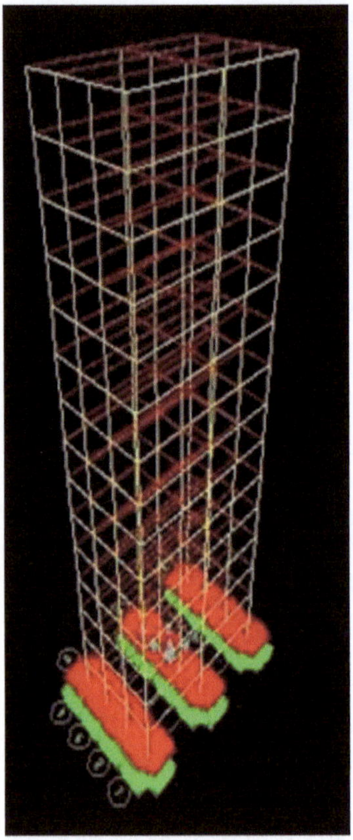

Fig. 6. Elevation of BF with isolated footing (modified winkler method)

(Fig. 3), Elevation of Bare frame with Fixed support shown in (Fig. 4), Elevation of Bare frame with Raft footing and Isolated footing by Modified Winkler method is shown in (Figs. 5 and 6) and also Elevation of Bare frame with Raft footing and Isolated footing by Continuum method is shown in (Figs. 7 and 8).

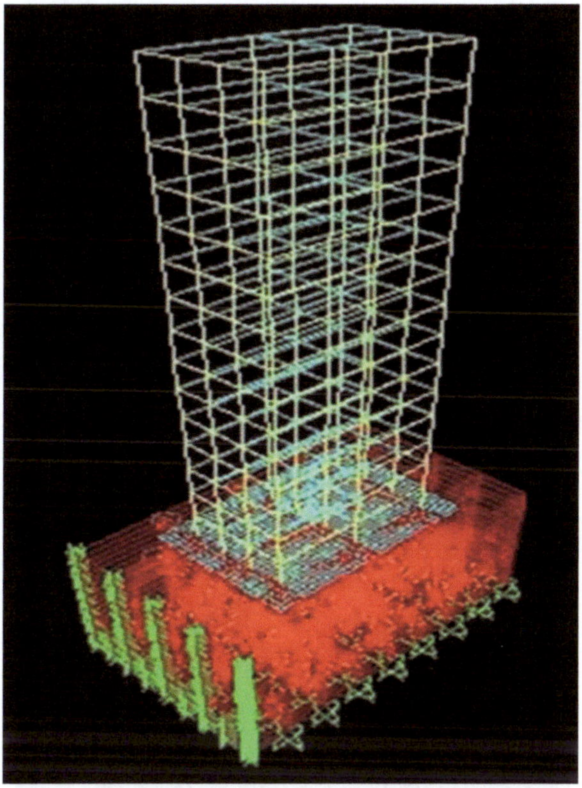

Fig. 7. Elevation of BF with raft footing (continuum method)

4 Result and Discussion

Results obtained from the analysis of 15 story framed structure for bare frame with rigid base and flexible base are tabulated for Vertical load and Lateral load. The maximum values obtained among all the load cases and load combinations considered are presented in Tables 2 3, 4, 5, 6 and 7. The discussion focus on the comparison between the two methods of flexible base with commonly practiced rigid base having following parameters.

a. Max. Joint Displacement in X & Y direction (Table 2)
b. Max. Time Period (Table 3)
c. Max. Column Forces (Tables 4 and 5)
d. Max. Support reaction (Tables 6 and 7)

4.1 Max. Joint Displacement

Maximum joint displacement in X and Y direction with corresponding load cases is given in Table 2.

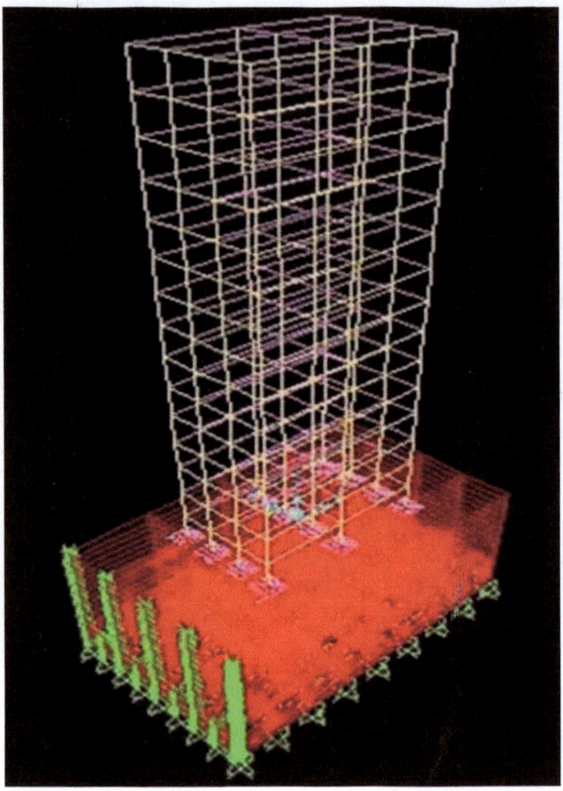

Fig. 8. Elevation of BF with isolated footing (continuum method)

Max. Displacement is one of the important parameter to decide the behavior of the structure. In the present case Max. Displacement in X direction is observed in Modified winkler method whereas there is not much difference in Continuum and Rigid base. The Raft footing shows lesser displacement compared to Isolated footing due to larger area and mass of footing. Also the Max. Displacement in Y direction is more compared to X direction this may be because of Moment of Inertia of column which is greater in X direction. It is observed that the soft soil shows Max. Displacement compared to other types of soil which indicates the necessity of analyzing building for soil structure interaction. Though there is not much difference in Continuum and Rigid base, Continuum method can be the best method chosen for future design of structure as the soil properties are taken into consideration and also it is modeled as 3D finite element model having the soil spread on sides and also beneath the structure. With respect to all Load combinations considered maximum displacement in X-direction occurs in load case 10 i.e. 1.5[DL + SQX] while in Y-direction in Load case 12 & 16 i.e. 1.5 [DL + SQY] & 0.9DL + 1.5SQY which shows the necessity and importance of load combinations in analysis.

Table 2. Max. Displacement for both Raft footing and Isolated footing

Raft footing	Load Case	X-dir in mm	Load Case	Y-dir in mm	Isolated footing	Load Case	X-dir mm	Load Case	Y-dir in mm
Rigid support									
Hard soil	1.5(DL + SQX)	0.91	1.5(DL + SQY)	2.04	Hard	1.5(DL + SQX)	0.91	0.9DL + 1.5SQY	2.04
Medium	1.5(DL + SQX)	1.24	1.5(DL + SQY)	2.78	Medium	1.5(DL + SQX)	1.24	1.5(DL + SQY)	2.78
Soft soil	1.5(DL + SQX)	1.52	1.5(DL + SQY)	3.41	Soft soil	1.5(DL + SQX)	1.52	1.5(DL + SQY)	3.41
Modified winkler method									
Hard soil	1.5(DL + SQX)	1.05	0.9(DL + 1.5SQY)	2.36	Hard	1.5(DL + SQX)	0.95	0.9DL + 1.5SQY	2.14
Medium	1.5(DL + SQX)	1.43	1.5(DL + SQY)	3.21	Medium	1.5(DL + SQX)	1.30	1.5(DL + SQY)	2.90
Soft soil	1.5(DL + SQX)	1.76	1.5(DL + SQY)	3.94	Soft soil	1.5(DL + SQX)	1.59	1.5(DL + SQY)	3.57
Continuum method									
Hard soil	1.5(DL + SQX)	0.92	1.5(DL + SQY)	2.39	Hard	1.5(DL + SQX)	0.92	1.5(DL + SQY)	2.05
Medium	1.5(DL + SQX)	1.26	1.5(DL + SQY)	3.27	Medium	1.5(DL + SQX)	1.27	1.5(DL + SQY)	2.82
Soft soil	1.5(DL + SQX)	1.58	1.5(DL + SQY)	4.07	Soft soil	1.5(DL + SQX)	1.60	0.9DL + 1.5SQY	3.51

4.2 Max. Time Period

Maximum Time Period analyzed is given in Table 3.

Table 3. Max. Time Period in secs, Frequency in cyc/sec and Eigen value in rad2/sec

Rigid support								
Raft footing	Mode no.	Period in sec	Frequency in cyc/sec	Eigen value rad2/sec	Isolated footing	Period in sec	Frequency in cyc/sec	Eigen value rad2/sec
Hard soil	1	8.97	0.11	0.49	Hard soil	8.97	0.11	0.49
Medium soil	1	8.97	0.11	0.49	Medium	8.97	0.11	0.49
Soft soil	1	8.97	0.11	0.49	Soft soil	8.97	0.11	0.49
Modified winkler method								
Hard soil	1	8.97	0.11	0.49	Hard soil	8.97	0.11	0.49
Medium soil	1	8.97	0.11	0.49	Medium	8.97	0.11	0.49
Soft soil	1	8.97	0.11	0.49	Soft soil	8.97	0.11	0.49
Continuum method								
Hard soil	1	9.69	0.1	0.42	Hard soil	8.98	0.11	0.43
Medium soil	1	9.73	0.65	0.42	Medium	9.02	0.11	0.49
Soft soil	1	9.78	0.64	0.41	Soft soil	9.08	0.11	0.48

Stiffness and mass of building along with its height plays an important role for a building to behave with larger or shorter time period. However when the structure is analysed for soil structure interaction the soil is idealized by many methods and the modulus of soil is mainly responsible for stiffness of the structure. As the stiffness of the structure reduces the natural period of structure increases. Compared to all the three types of soil the stiffness of the structure in soft soil is less as of which the natural period of the structure increases in soft soil. Also since the continuum method is modeled as 3D soil model with all the properties of soil input in the model, the continuum method shows realistic results. The structure analysed by Continuum method experiences increase in time period as stiffness of structure decreases which indicates that as the structure is more flexible they have larger time period.

4.3 Max. Column forces

Maximum Column forces analyzed is given in Tables 4 and 5.

Axial Stiffness of the column is mainly responsible for Lateral sway of the building. From the analysis result it is observed that as axial load is larger in soft soil for all three methods of analysis bending moment increases, this behavior can be taken as indication to increase the cross section area of column in case of soft soil.

Table 4. Max. Column Forces –Raft Footing

Rigid support						
Raft footing	Load case	Max axial force Fx KN	Load case	Max moment My KNm	Load case	Max moment Mz KNm
Hard soil	SQY	1738.87	1.5(DL + SQY)	539.1	1.5(DL + SQX)	673.67
Medium	SQY	2364.87	1.5(DL + SQY)	732.69	1.5(DL + SQX)	916.18
Soft soil	SQY	2903.92	1.5(DL + SQY)	899.4	1.5(DL + SQX)	1125.02
Modified winkler method						
Hard soil	SQY	2005.38	0.9DL + 1.5SQY	620.98	1.5(DL + SQX)	776.9
Medium	SQY	2727.32	1.5(DL + SQY)	844.79	1.5(DL + SQX)	1056.58
Soft soil	SQY	3348.98	1.5(DL + SQY)	1037.04	1.5(DL + SQX)	1297.42
Continuum method						
Hard soil	SQY	1685.18	1.5(DL + SQY)	565.3	1.5(DL + SQX)	690.26
Medium	SQY	2287.07	1.5(DL + SQY)	771.13	1.5(DL + SQX)	953.25
Soft soil	SQY	2806.25	1.5(DL + SQY)	946.89	1.5(DL + SQX)	1167.19

Table 5. Max. Column Forces – Isolated Footing

Rigid support						
Isolated footing	Load case	Axial force Fx KN	Load case	Moment My KNm	Load case	Moment Mz KNm
Hard soil	SQY	1738.87	1.5(DL + SQY)	539.1	1.5(DL + SQX)	673.65
Medium	SQY	2364.87	1.5(DL + SQY)	732.7	1.5(DL + SQX)	916.17
Soft soil	SQY	2903.92	1.5(DL + SQY)	899.4	1.5(DL + SQX)	1125
Modified winkler method						
Hard soil	SQY	1816.29	1.5(DL + SQY)	563.04	1.5(DL + SQX)	703.64
Medium	SQY	2470.15	1.5(DL + SQY)	765.25	1.5(DL + SQX)	956.96
Soft soil	SQY	3033.2	1.5(DL + SQY)	939.38	1.5(DL + SQX)	1175.08
Continuum method						
Hard soil	SQY	1737.56	1.5(DL + SQY)	543.78	0.9DL + 1.5SQX	674.77
Medium	SQY	2360.51	1.5(DL + SQY)	760.41	1.5(DL + SQX)	919.45
Soft soil	SQY	2898.6	1.5(DL + SQY)	965.85	1.5(DL + SQX)	1135.17

4.4 Max. Support Reaction

Maximum Support reaction analyzed is given in Tables 6 and 7.

The column base fixity is responsible for the overall behavior of structure. It is observed in the analysis that continuum method shows more variation in response compared to other method this may be due to soil flexibility which allows the column base to behave as hinge with lesser rotation which is responsible for lesser lateral displacement.

Table 6. Max. Support Reaction - Raft footing

Raft footing	Load case	Max Fz KN	Load case	Max moment Mx KNm	Load case	Max moment My KNm
Rigid support						
Hard soil	1.5(DL + SQY)	7754.33	1.5(DL + SQY)	539.098	1.5(DL + SQX)	673.669
Medium	1.5(DL + SQY)	8693.32	1.5(DL + SQY)	732.693	1.5(DL + SQX)	916.182
Soft soil	1.5(DL + SQY)	9501.9	1.5(DL + SQY)	899.401	1.5(DL + SQX)	1125.02
Modified winkler method						
Hard soil	1.5(DL + LL)	7707.42	0.9DL + 1.5SQY	620.99	1.5(DL + SQX)	776.9
Medium	1.5(DL + SQY)	9469.33	1.5(DL + SQY)	844.82	1.5(DL + SQX)	1056.58
Soft soil	1.5(DL + SQY)	10583.7	1.5(DL + SQY)	1037.18	1.5(DL + SQX)	1297.42
Continuum method						
Hard soil	1.5(DL + LL)	1782.01	1.5(DL + SQX)	60.72	1.5(DL + LL)	50.11
Medium	1.5(DL + SQY)	1489.22	1.5(DL + SQX)	76.98	1.5(DL + SQX)	33.39
Soft soil	1.5(DL + SQY)	1680.26	1.5(DL + SQY)	83.08	1.5(DL + SQX)	18.47

Table 7. Max. Support Reaction - Isolated footing

Isolated footing	Load case	Fz KN	Load case	Moment Mx KNm	Load case	Moment My KNm
Rigid support						
Hard soil	1.5(DL + SQY)	7757.84	1.5(DL + SQY)	539.098	1.5(DL + SQX)	673.653
Medium	1.5(DL + SQY)	8696.84	1.5(DL + SQY)	732.695	1.5(DL + SQX)	916.169
Soft soil	1.5(DL + SQY)	9505.41	1.5(DL + SQY)	899.401	1.5(DL + SQX)	1125
Modified winkler method						
Hard soil	1.5(DL + SQY)	8005.65	1.5(DL + SQY)	563.039	1.5(DL + SQX)	703.643
Medium	1.5(DL + SQY)	8986.44	1.5(DL + SQY)	765.253	1.5(DL + SQX)	956.955
Soft soil	1.5(DL + SQY)	9831.02	1.5(DL + SQY)	939.382	1.5(DL + SQX)	1175.09
Continuum method						
Hard soil	1.5(DL + LL)	1632.24	1.5(DL + SQX)	81.362	1.5(DL + SQX)	54.278
Medium	1.5(DL + LL)	1634.89	1.5(DL + SQX)	169.09	1.5(DL + SQX)	28.19
Soft soil	1.5(DL + LL)	1803.21	1.5(DL + SQY)	200.158	1.5(DL + SQX)	16.626

5 Conclusions

1. Max. Displacement is a major factor which predicts the behavior of structure the results shows that though there is not much difference between Continuum method and rigid base still we can consider Continuum method as realistic method for future design of structures as soil is modeled as 3D model by specifying soil properties for different types of soil, also continuity of soil is taken into account by considering soil around and beneath the structure.
2. Max. Displacement is less in Continuum method compared to Modified Winkler method which also indicates Continuum method as realistic method.
3. Max. Displacement in Y direction is greater than X direction this is because the Moment of Inertia of columns along X-direction is higher than Y direction.
4. Time period increases in Continuum method compared to Modified winkler method and rigid base as foundation soil system is rendered flexible.

5. Comparing types of footing Raft footing shows better results.
6. Axial force in Column is reduced and also there is a considerable difference in Support reaction in continuum method for both Raft footing and Isolated footing this may be due to stiffness of soil which gives an indication for change of column size in soft soil.

However much more research work has to be conducted on Continuum method based on the length, width and depth of soil conditions along with boundary conditions.

References

1. Garag, V., Karuna, S.: Effect of soil-structure interaction on regular and irregular, medium RC framed structure. Int. J. Sci. Eng. Technol. Res. (IJSETR) **5**(5) (2016)
2. Patil, S.S., Kalyanshetti, M.G., Dyawarkonda, S.S.: Parametric study of R.C frames with raft foundation considering soil structure interaction using spring. Int. J. Sci. Develop. Res. (IJSDR) **1**(4), pp. 63–67 (2016)
3. Mathew, A.K., Tomer, S.K., Lovely, K.M.: Effect of soil-structure interaction in seismic analysis of framed structures using ansys. Int. J. Eng. Develop. Res. IJEDR **3**(3), 1–9 (2015)
4. Bajad, M.N., Sawant, R.: Effect of soil structure interaction on high rise RC building. IOSR J. Mech. Civil Eng. **13**(1), 85–91 (2016)
5. Chougule, A.R., Dyavanal, S.S.: Seismic soil structure interaction of buildings with rigid and flexible foundation. Int. J. Sci. Res. ISSN **4**(6) (2015)
6. Zubair, M., Shilpa, B.R.: A parametric study of soil structure interaction of raft foundation by using dynamic analysis. Int. J. Eng. Sci. Invent. Res. Develop. **3**(1) (2016). e-ISSN: 2349-6185
7. Ismail, A.: Effect of soil flexibility on seismic performance of 3-D frames. IOSR J. Mech. Civil Eng. IOSR-JMCE **11**(4), 135–143 (2014)
8. Bhat, R., Warad, S.A.: Seismic evaluation of RC building considering SSI. IJS&T **3**(11) (2015)

Geotechnical Research and Design of Wind Power Plant

R. E. Lukpanov(✉), Talal Awwad, D. K. Orazova, and D.V Tsigulyov

Eurasian National University of L.N. Gumilyov, Astana, Kazakhstan
rauan_82@mail.ru

Abstract. The paper refers to one of the sought-after directions – sustainable development and renewable energy. According to the world practice, the wind power industry has received a powerful impetus for development and has reached the leading positions in the economies of developed countries.

The paper presents the results of the research of the wind power plant foundations. The studies included in-situ observations (monitoring) of wind speed and vibration of the base plate (slab) of the wind power plant. Comparison of the results of the intensity of the wind pressure and the vibration parameters of the foundation helped to assess the vibration effect on the foundation in the windiest period (according to the standard for construction climatology).

After assessing the reliability of the operated wind power plant based on the results of in-situ observations, the output data of the maximum values of the parameters were obtained. These parameters had used for numerical analysis in the software Plaxis 2D to compare the data of the dynamic impact on the foundation from the tower of the power plant. Numerical simulation of the foundation of the wind power plant was carried out with the aim of analyzing the stress-strain state of the soil massif and evaluation of the stability of the basement.

1 Introduction

A lot of studies have been done to analyze the dynamic behavior of soils (Awwad and Donia 2016; Awwad et al. 2017; Alkayyal and Awwad 2015). However, there are not a lot of publication about the wind power plant foundations, especially for the geotechnical conditions of Kazakhstan. Wind power is the most dynamically developing type of renewable energy sources. Having studied the energy potential of wind in Kazakhstan, the Government of the Republic of Kazakhstan together with the UN development Program "Kazakhstan is the initiative of the development of wind power market" has resolved that the Ereymentau district of Akmola region is the most prospective area for the construction of wind power plants (National program 2007).

The research project represents the Wind Power Plant (WPP hereinafter) with the capacity of 50 MW located near Ereymentau city (Kazakhstan). The WPP is under construction and provided for increasing wind energy power in Ereymentau up to 300 MW. WPP located in a high wind zone, and this makes it useful for large-scale production of electric power.

D. Choudhury et al. (Eds.): GeoMEast 2018, SUCI, pp. 220–227, 2019.
https://doi.org/10.1007/978-3-030-01920-4_19

The construction site is located on the undulating foothill plain, which has the general slope to the north. The geological structure of this territory has the sedimentary and metamorphic rocks of the Proterozoic and Paleozoic era, which are broken by intrusions in the north-east part of the city and covered with eluvia-diluvia quaternary deposits. The geological cross section is presented in Fig. 1.

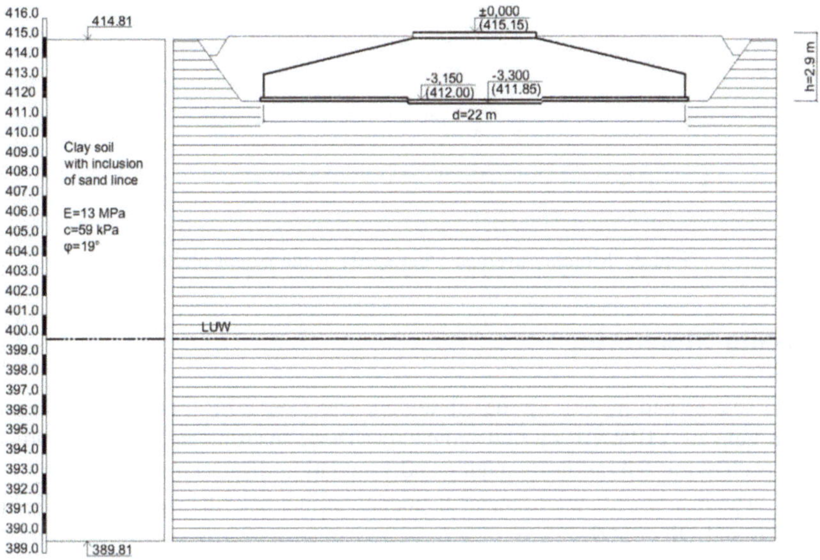

Fig. 1. Geological section of the site

Annual measurements of speed and direction of wind were made at the site within the context of the UNDP project on wind power. Measurements were performed in accordance with international standards in the field of measuring wind speed in order to estimate wind potential (IEA/IEC) [2].

The distribution of wind speed and Weibull parameters at the height of 51 m (the axis of the gondola) for the site of Ereymentau WPP are shown in Fig. 2.

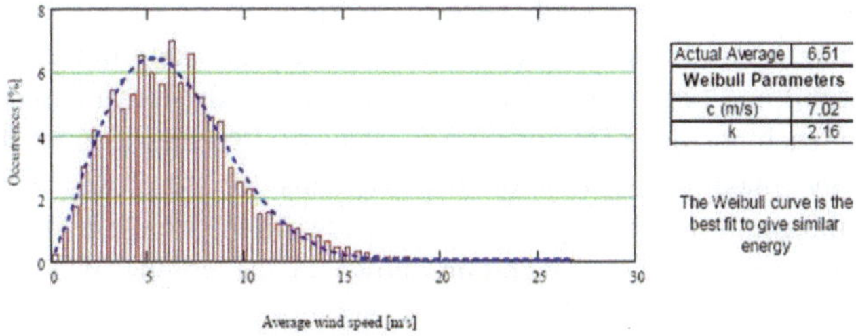

Fig. 2. Distribution of wind speed and Weibull parameters at the height of 51 m

According to the results of wind pressure measurements, the diagram of a seasonal distribution of wind speed was made. It demonstrates the changes in the wind flow speed by month in relation to average annual wind speed (Fig. 3).

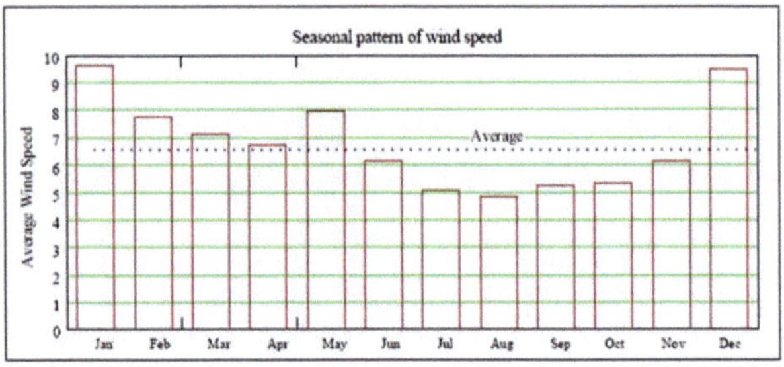

Fig. 3. Monthly average wind speed at the height of 50 m

2 In-Situ Monitoring

The field studies and measuring the vibrational effect on WPP foundation were made with the help of VIBRA Profound indicators. These instruments allow defining the speed, acceleration, vibration frequency and foundation displacement caused by wind pressure. The instrument system meets national and international standards SBR 2002, DIN 4150 and DIN 45669.

The source of vibration is the WPP tower located on a reinforced concrete foundation with a thickness of 2.9 m and a diameter of 22 m. Fluctuations caused by the tower, are passed to the ground base through the foundation Zhussupbekov et al. (2017).

Vibration loads from the tower to the foundation were measured by sensors with the help of VIBRA Profound device (Fig. 4). Movements are recorded on the WPP tower (Fig. 4). Intermediary measurements were counted every 10 s.

A measurement takes a month to make all the measurements. Control points data was obtained according to the diagrams of the device. The relationship between wind pressure and vibration impacts on the WPU foundation from the tower was obtained after data analyzing. The method of extrapolation can allow measuring the impact on the foundation at a maximum wind speed of 25 m/s.

The diagram in Fig. 5 shows a dependence of the foundation movement (due to of tower fluctuation) and time elapses. The diagram shows the maximum values of displacements for each position. The maximum values of displacements for the 1 position are 0,1 mm, the maximum values of displacements for the 2 and 3 positions are 0,2 mm, and 0,3 mm – for the 4 position, at the maximum wind pressure of 4,75 m/s.

The measurement results of vibration shown in the graphs of a dependence of displacements, accelerations, frequencies of vibration, a velocity of vibration from wind speed (Fig. 6).

According to the results, the measurement was made the prediction of vibration parameters of the wind turbine foundation from wind pressure.

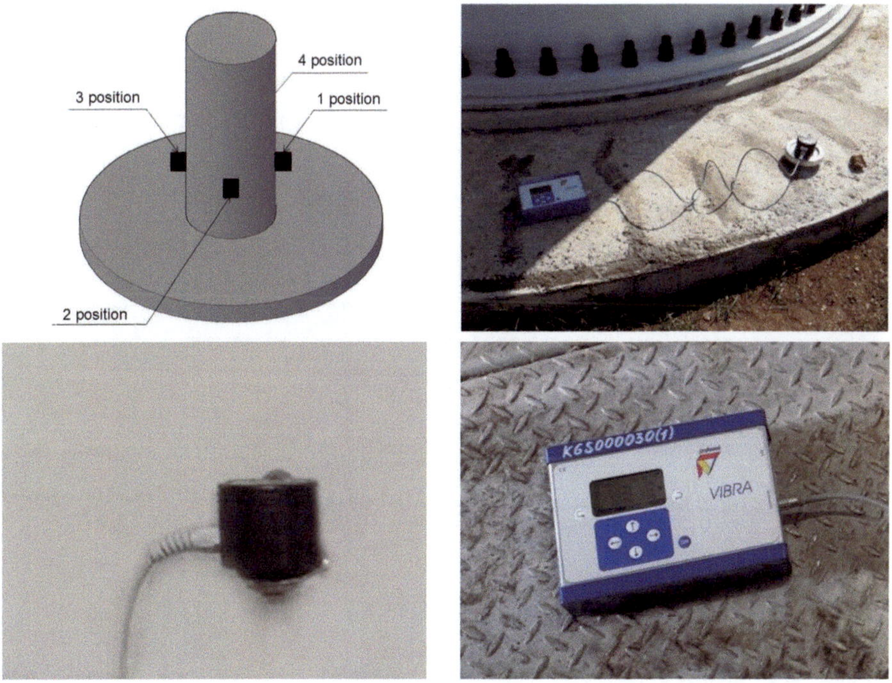

Fig. 4. In-situ monitoring of wind power

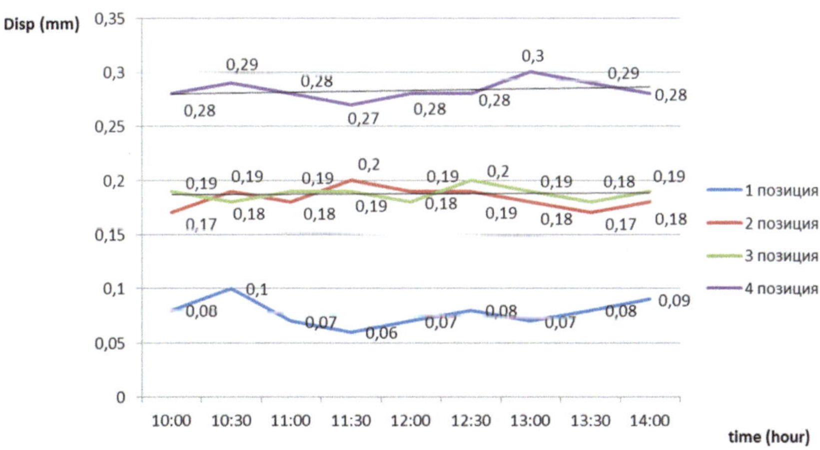

Fig. 5. Diagram of a dependence of displacement and time

In accordance of field measurement maximum values of the parameters at the maximum wind pressure in this region are follows:

(1) Maximum displacement on X = 0,038 mm, Y = 0,031 mm, Z = 0,08 mm;

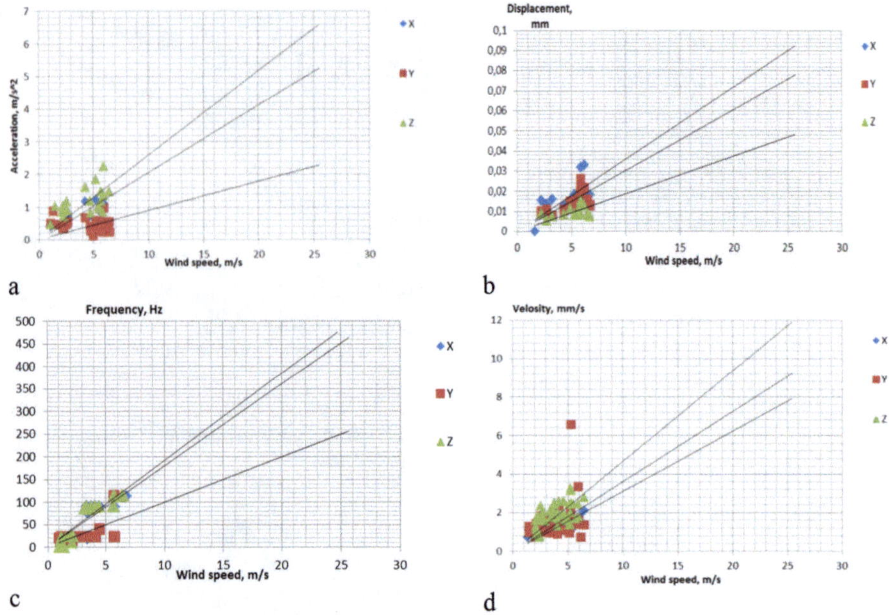

a) graphs of a dependence of displacements from wind speed
b) graphs of dependence frequencies of vibration from wind speed
c) graphs of a dependence of accelerations from wind speed
d) graphs of a dependence of velocity of vibration from wind speed

Fig. 6. Monitoring data

(2) Maximum acceleration on X = 2,8 m/s^2, Y = 2,1 m/s^2, Z = 0,7 m/s^2;
(3) Maximum frequencies of vibration on X = 122 Hz, Y = 120 Hz, Z = 65 Hz;
(4) Maximum velocity of vibration on X = 4,2 mm/s, Y = 3,6 mm/s, Z = 2,8 mm/s.

3 Numerical Modeling of WPP

PLAXIS 2D allows modelling the interaction between soil and construction. The influence of vibration work on the soil on which it is installed was considered.

The dimensions of the soil mass are determined based on the computation of compressible stratum, represented by a clay layer of 60 m. The soil surface is set at the point y = 0. The foundation is represented by clay, which is considered here as an elastic-plastic material. A dataset is formed by selecting the soil type and interface, according to the information given in Table 1. Dynamic stiffness of soil is significantly higher than its static stiffness, as dynamic loading is usually rapid and it results in ultra-small deformations.

Table 1. Material properties of soil

Parameter	Name	Value	Unit
General			
Material model	Model	Mohr-Coulomb	–
Type of material behaviour	Type	Drained	
Soil unit weight above the phreatic level	γ_{unsat}	16	kN/m^3
Soil unit weight below the phreatic level	γ_{sat}	18	kN/m^3
Parameters			
Young's modulus (constant)	E'	13*10^4	kN/m^2
Poisson's ratio	ν'	0,38	
Initial			–
K_o determination	–	Automatic	–
Lateral earth pressure coefficient	$K_{o,x}$	0,674	–

The material dataset for footing according to the information given in Table 2.

Table 2. Material properties of the footing

Parameter	Name	Value	Unit
General			
Material model	Model	Linear elastic	–
Type of material behaviour	Type	Non–porous	–
Soil unit weight above the phreatic level	γ_{unsat}	24	кН/м^3
Parameters			
Young's modulus (constant)	E'	30*10^6	кН/м^2
Poisson's ratio	ν'	0,2	
Initial			–
K_o determination	–	Automatic	–
Lateral earth pressure coefficient	$K_{o,x}$	1	–

A distributed load is applied to the foundation, as well as a horizontal force and a concentrated moment for modelling the wind turbine weight, and fluctuations that it causes Lukpanov et al. (2016). The geometric model is shown in Fig. 7.

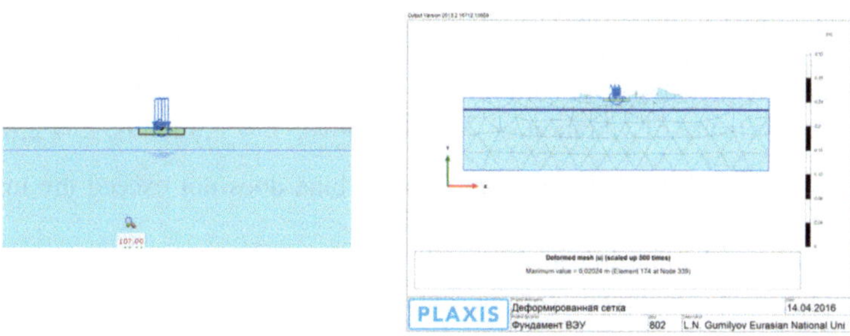

Fig. 7. Geometric model and mesh of the foundation

Figures 8, 9 and 10 show the results of the accelerations, the results velocity of vibration values and the diagram of the dependence of vertical displacement on time.

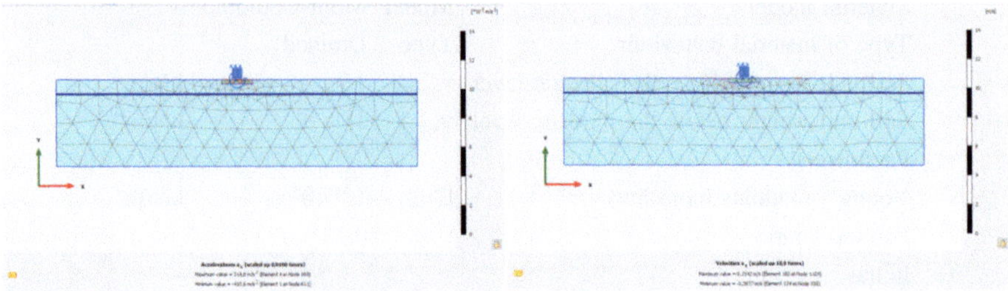

Fig. 8. The results of the acceleration and velocity

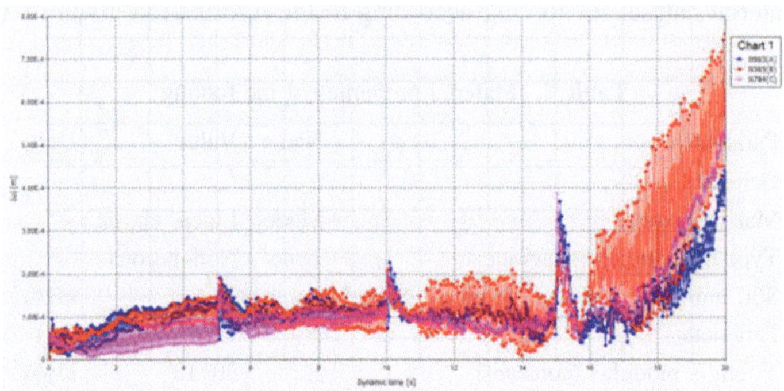

Fig. 9. The diagram of the dependence of vertical displacement on time

4 Conclusions

According to the results, the measurement was made the prediction of vibration parameters of the wind turbine foundation from wind pressure.

Maximum values of the parameters at the maximum wind pressure in this region are:

- Maximum displacement on X = 0,038 mm, Y = 0,031 mm, Z = 0,08 mm;
- Maximum acceleration on X = 2,8 m/s2, Y = 2,1 m/s2, Z = 0,7 m/s2;
- Maximum frequencies of vibration on X = 122 Hz, Y = 120 Hz, Z = 65 Hz;
- Maximum velocity of vibration on X = 4,2 mm/s, Y = 3,6 mm/s, Z = 2,8 mm/s.

This parameter of displacements for wind turbine does not exceed the maximum permissible value of 5 mm (DIN 45669-1 1995).

The obtained vibration impacts from the tower to the foundation under maximum wind pressure were integrated with the Plaxis 2D software program. Having applied static and dynamic loads in the program we got the displacement with a certain frequency and acceleration, and vibrations speed.

The diagram of the dependence of vertical displacement on time shows that these displacements do not exceed a limit value for WPU foundations.

References

Alkayyal, H., Awwad, T.: Improvement of the dynamic behavior of soil structures underlain by liquefiable soil using the Geosynthetics-encased columns. In: Proceedings of the 6th International Symposium on Deformation Characteristics of Geomaterials, Buenos Aires, Argentina, November 2015, Advances in Soil Mechanics and Geotechnical Engineering, vol. 6, pp. 792–799. IOS press (2015)

Project of the Government of Kazakhstan and the Development Program of the United Nations Organization "Kazakhstan - Initiative of the Development of Wind Power Engineering Market", the wind power plant near Ereymentau city, Pre-investing research, Almaty, pp. 7–9 (2008)

DIN 45669. Human exposure to vibration in buildings, Germany, p. 8 (1995–2006)

Zhussupbekov, A. Zh., Lukpanov, R.E., Tsygulyov, D.V., Orazova, D.K.: Research of vibrating influence of wind power tower to the foundation. Вестник НИА РК №4 (66), pp. 124–130 (2017)

Lukpanov, R.E., Orazova, D.K., Zhussupbekov, A., Zh., Sapenova, Zh.: Design of foundation for wind turbine with analysis by finite element method. In: 13th Baltic Sea Geotechnical Conference, Lithuania, 22–24 September 2016, pp. 196–200 (2016). ISSN 2424-5968, ISBN 978-609-457-957-8

Awwad, Talal, et al.: Effect of a stiff thin Foundation soil layer's depth on dynamic response of an embankment dam. Procedia Eng. **189**(2017), 525–532 (2017)

Awwad, T., Donia, M.: The efficiency of using a seismic base isolation system for a 2D concrete frame founded upon improved soft soil with rigid inclusions. Earthquake Eng. Eng. **15**(1), 49–60 (2016). Institute of Engineering Mechanics

The national program of wind energy development until 2015 with the perspective till 2024 (project), Almaty, Astana, pp. 3–5 (2007)

Author Index

D. Choudhury et al. (Eds.): GeoMEast 2018, SUCI, pp. 229–230, 2019.
https://doi.org/10.1007/978-3-030-01920-4

MIX
Papier aus verantwortungsvollen Quellen
Paper from responsible sources
FSC® C105338

If you have any concerns about our products,
you can contact us on
ProductSafety@springernature.com

In case Publisher is established outside the EU,
the EU authorized representative is:
Springer Nature Customer Service Center GmbH
Europaplatz 3, 69115 Heidelberg, Germany

Printed by Libri Plureos GmbH
in Hamburg, Germany